PHP Web Development with Laminas

Build a fully secured and functional e-commerce application
with PHP using the Laminas framework

Flávio Gomes da Silva Lisboa

BIRMINGHAM—MUMBAI

PHP Web Development with Laminas

Group Product Manager: Pavan Ramchandani

Publishing Product Manager: Kushal Dave

Senior Editor: Mark D'Souza

Senior Content Development Editor: Rakhi Patel

Technical Editor: Simran Udasi

Copy Editor: Safis Editing

Project Coordinator: Sonam Pandey

Proofreader: Safis Editing

Indexer: Sejal Dsilva

Production Designer: Nilesh Mohite

Marketing Coordinators: Anamika Singh and Marylou De Mello

First published: November 2022

Production reference: 1201022

Published by Packt Publishing Ltd.
Livery Place
35 Livery Street
Birmingham
B3 2PB, UK.

ISBN 978-1-80324-536-2

www.packt.com

To my wife, Maria, a warrior, the love of my life, who beat cancer. To my daughter, Koriander, a superheroine, who has empathy as one of her superpowers. To the memory of my mother, Cleonisse, who taught me how to read and let me travel the world in books.

– Flávio Lisboa

Foreword

PHP Web Development with Laminas by Flávio Gomes da Silva Lisboa is a very good book if you are interested in web development with PHP. The author shows how to implement, using a step-by-step approach, a complete e-commerce application with the Laminas open source project (formerly Zend Framework). I really liked the approach of the author, introducing not only the usage of the framework but also many best practices such as test-driven development, behavior-driven development, principles of Agile development, and more. This is definitely a book to have if you want to build professional web applications in PHP.

Enrico Zimuel, principal software engineer at Elastic Core Committee, member of the PHP Framework Interop Group, formerly Zend Framework core developer

Contributors

About the author

Flávio Gomes da Silva Lisboa is a development analyst in SERPRO involved in several projects for the Brazilian government, an instructor of development software courses, a teacher of development software disciplines, and a member of the Brazilian PHP community. He also was a programmer of desktop applications for restaurants and industrial kitchens and a system analyst on the international board of the Bank of Brazil. Sometimes, he writes sci-fi romances such as *The One: The Solitude and the Harmony* and *Pandino, the Emperor*. You can read more about him and his adventures at `https://fgsl.eti.br`.

I want to thank the people who have been close to me and supported me, especially my wife Maria and my daughter Koriander. They are the reason for which I want to progress. This text is readable thanks to the uncanny work of Divya Selvaraj and Rakhi Patel. I also need to thank Er Galvão and Matt Fletcher for their accurate technical reviewing. I can't forget Eduardo Bona and Anderson Burnes, who were course coordinators when I taught Zend Framework, now Laminas, to several classes of IT college students. It is necessary to also thank Marcos Melo, Walter Zapalowski, Guilherme Funchal, and Antonio Tiboni, fellows in SERPRO, who helped me to work on an international project that used Zend Framework. Finally, I need to mention the years of partnership with Anderson de Paula in training PHP developers. Of course, this book wouldn't be possible without the wonderful support of the Packt Publishing team, which gave me an incredible experience of writing. Thank you very much to Kushal Dave, Mark D'Souza, Sonam Pandey, and every member of Packt Publishing who was a part of the building of this book.

About the reviewer

Matt Fletcher is an experienced software engineer who has designed and developed PHP web applications for over 20 years. He is currently the development manager for SilverDoor Apartments, the leading booking agent for serviced apartments worldwide. He leads an international team of developers who use the Laminas framework and API Tools to deliver a range of enterprise tools. His work includes reviewing pull requests, planning system architectures, and delivering high-quality projects.

Table of Contents

Preface   xiii

Part 1: Technical Background

1

Introducing Laminas for PHP Applications   3

Why do we need web frameworks?   3   Zend Framework – the reference architecture
Continuous integration and continuous delivery 4   for PHP applications   8
The usefulness of a framework   6   The differences between Zend Framework
                                                       and Laminas   10
From Zend Framework to Laminas   6
My introduction to Zend Framework   7   The technical and social
                                                       infrastructure of the Laminas
                                                       community   10
                                                       Summary   12

2

Setting Up the Environment for Our E-Commerce Application   13

Technical requirements   14   Configuring the IDE – Eclipse for
Installing Laminas   15   PHP developers   18
Installing Laminas through an isolated   Integrating Laminas and Eclipse   21
component of Laminas   16   Managing MySQL   29
Installing Laminas through an MVC   Summary   34
skeleton application   16

3

Using Laminas as a Library with Test-Driven Development   37

Technical requirements	37	Configuring the autoloading of test classes	57
Understanding the software requirements	38	Creating the config.php file and rerunning the SchoolClassTest test case	58
Use case – managing school classes	38	Creating a new testInserting test method	60
Implementing a structure for the use case	39	Creating an edition form edit.php	63
Creating our first automated test	45	Creating a SchoolClassTest->testUpdating test method	65
Developing our use case with TDD	49	Altering edit.php	68
Creating our SchoolClassTest test class	50	Altering save.php	70
Creating a test method in the SchoolClass class	53	Creating the delete.php file	70
Creating the SchoolClass class	54	Generating code coverage reports	71
Configuring autoloading of classes	55	Summary	74

4

From Object-Relational Mapping to MVC Containers   75

Technical requirements	75	Using Laminas as a container	92
Using ORM	76	Creating the school3 project	92
Use case – managing students	78	Creating the School module	93
Implementing the use case	78	Ensuring the module has the right namespace	95
Running the SchoolClassTest->testListing test	80	Implementing the test methods	96
Adjusting CRUD files	87	Running the tests	97
		Summary	108

Part 2: Creating an E-Commerce Application

5

Creating the Virtual Store Project   111

Technical requirements	111	Understanding the project requirements for the sample web application	112

Identifying use cases and actors   113

Understanding the class diagram
for Whatstore   115

Creating the project instance   116

Understanding the structure of a
Laminas project   117

Preparing the modules for our project   118

Summary   122

6

Models and Object-Relational Mappers with Behavior-Driven Development   123

Technical requirements   123

BDD with Behat   124

Creating our first scenario   126

Mapping a user story to an automated test   127

Creating models from user stories   131

Creating mappers from user stories   135

Adding a product to the products table   136

Checking whether the product was inserted   140

Creating a generic model and a
generic mapper   142

Creating other models with Behat   146

Summary   155

7

Request Control and Data View   157

Technical requirements   157

Understanding the relationship
between HTTP and PHP   157

The request life cycle in the
Laminas MVC   158

Implementing CRUD with a
controller and view pages   160

Testing our current test cases   160

Testing product insertion   164

Testing product recovery   167

Testing product updates   168

Testing product deletion   169

Creating the ProductAPIController class   170

Implementing the ProductController class   173

Implementing the pages for
ProductController   178

Implementing the interaction between the
web interface and API   182

Testing the interaction between the web
interface and API   188

Summary   193

8

Creating Forms and Implementing Filters and Validators   195

Technical requirements	195	Creating a class to define an HTML form	200
Generating a form for discounts with laminas-form	196	Filtering input data	205
		Validating input data	208
Creating CRUD implementation using the laminas-form component	196	Summary	212

9

Event-Driven Authentication   213

Technical requirements	213	Creating the employee's registration	226
Creating login and logout actions for the Inventory module	214	Testing the login page	228
Creating tests for login and logout actions	214	Verifying user identity based on events	232
Creating login and logout actions	215	Delegating the authentication listener to a module	238
Implementing the user authentication for the Inventory module	220	Summary	240

10

Event-Driven Authorization   241

Technical requirements	242	Implementing access control	259
Creating roles registration	242	Verifying users' permissions	263
Creating resources registration	243	Modifying the test bootstrap	265
Filling the resources table	244	Creating an identity manager	267
Associating resources with roles	250	Summary	275
Associating roles with users	254	Further reading	275

Part 3: Review and Refactoring

11

Implementing a Product Basket | 279

Technical requirements	279	Order closing	301
Managing the product inventory	280	Decoupling the generic behavior of authentication	305
Refactoring the customer home page	287		
Controlling the product basket	293	Summary	308
Controlling the customers	297		

12

Reviewing and Improving Our App | 309

Technical requirements	310	Generating API documentation	321
Principles of agile development	310	Isolating identity management for each module	329
Eliminating duplicated code	311		
Removing unnecessary configuration	316	Modifying the class for identity management	330
Creating a product search box	318	Refactoring listener and controller classes	332
		Summary	334

13

Tips and Tricks | 335

Technical requirements	336	Uploading files with laminas-form	345
Creating mapped models	336	Changing the application layout	348
Customizing filters and validators	337	Mastering the controller layer	350
Creating a customized filter	338	Detecting AJAX requests	350
Creating a customized validator	340	Exchanging messages between the controller and view layers	352
Understanding the Laminas view layer	342	Finishing controller actions	354
Producing JSON responses with laminas-view	343	Summary	354

14

Last Considerations   357

Other Laminas components   357
Cache   357
CLI   358
Crypt   358
CSRF   358
DOM   358
Escaper   359
Feed   359

Hydrator   360
JSON-RPC server   360
LDAP   360
Log   361

Microservice-oriented development 361
Building APIs   362
Community resources   362
Summary   362

Index   363

Other Books You May Enjoy   372

Preface

Laminas is a framework for building PHP applications. There are several frameworks for any programming language, and this is no different for PHP. So, why would you choose Laminas? Well, Erich Gamma, a renowned computer scientist and the principal designer of widely used tools such as JUnit and Eclipse, states that a framework is at the highest level of reuse for an object-oriented system. For this, the framework needs to be under the control of the developer. A framework shouldn't be heavyweight, seeking to do much for the developer or doing it in a way that the developer doesn't want. In addition, frameworks should always allow the developer to change how they do things – they must be flexible. A framework should not be pretentious and force the developer to use all of its components as if they were the definitive implementation for a given problem. A framework should be humble enough to recognize that the developer is the one who has the best solution for a specific problem and allow them to integrate their solution into the framework's generic components. Laminas is a framework that meets Erich Gamma's expectations. Born in 2005 as Zend Framework, Laminas has gradually achieved maturity in its architecture, providing a collection of loosely coupled and configurable components for building object-oriented PHP applications.

The Laminas framework enables PHP web developers to create powerful web applications with an evolutionary architecture. Developers would want to learn Laminas to build web applications based on the reuse of loosely coupled components. Developers working with PHP will be able to put their knowledge to work with this practical guide to Laminas. This book provides a hands-on approach to implementation and associated methodologies that will have you up and running and being productive in no time. Complete with step-by-step explanations of essential concepts, practical examples, and self-assessment questions, you will be able to quickly create PHP programs reusing Laminas components.

You'll learn how to create the basic structure of a PHP web application divided into layers, understand the MVC components of Laminas, and understand how you can take advantage of the Eclipse platform when developing with Laminas. You'll then build an e-commerce application based on the requirements of a fictional business and learn how to implement these requirements with Laminas components. By the end of this book, you will be able to build an MVC application in the PHP language using Laminas.

Who this book is for

This book is aimed at PHP developers with basic knowledge of PHP fundamentals. It is for those developers starting their journey through web development with PHP, and wanting to learn how to work with the best tools and practices.

What this book covers

Chapter 1, Introducing Laminas for PHP Applications, presents the evolution of the architecture of Zend Framework up until it changed to Laminas, and discusses the lessons learned.

Chapter 2, Setting Up the Environment for Our E-Commerce Application, sees us set up the development environment for the example application that we will be building throughout this book.

Chapter 3, Using Laminas as a Library with Test-Driven Development, explains, using a simple example based on a reduced use case, how it is possible to use only one Laminas component in an existing PHP application.

Chapter 4, From Object-Relational Mapping to MVC Containers, covers the handling of databases with object-relational mapping, abstracting SQL statements, and the inversion of control for object-oriented PHP applications with the Laminas framework as a container.

Chapter 5, Creating the Virtual Store Project, presents the requirements of the e-commerce project we plan to build.

Chapter 6, Models and Object-Relational Mappers with Behavior-Driven Development, is where we use `laminas-db` for mapping the entity-relationship model to an object model. We also learn how to create models using the behavior-driven development approach.

Chapter 7, Request Control and Data View, shows how to use the Laminas implementations for the controller and view layers in an MVC application.

Chapter 8, Creating Forms and Implementing Filters and Validators, introduces some Laminas components for handling data input.

Chapter 9, Event-Driven Authentication, covers one of the essential topics for developing secure applications: authentication.

Chapter 10, Event-Driven Authorization, covers another one of the essential topics for developing secure applications: authorization.

Chapter 11, Implementing a Product Basket, describes the implementation of a product basket, an essential component of a virtual store.

Chapter 12, Reviewing and Improving Our App, is dedicated to reviewing and improving the sample e-commerce application.

Chapter 13, Tips and Tricks, provides some tips and tricks that will help you as a developer to solve some problems that can eventually appear when constructing a web application.

Chapter 14, Last Considerations, delivers the conclusion of the book, outlining the possibilities that lie ahead with other components not used in our sample applications.

To get the most out of this book

Software/hardware covered in the book	Operating system requirements
Laminas MVC 3.3.4 or above	Windows, macOS, or Linux
Eclipse IDE for PHP developers 4.24 or above	Windows, macOS, or Linux with JDK installed
XAMPP 7.4.29 / PHP 7.4.29 (minimal) or 8.1.6 / PHP 8.1.6 9 (recommended)	Windows, macOS, or Linux

If you are using the digital version of this book, we advise you to type the code yourself or access the code from the book's GitHub repository (a link is available in the next section). Doing so will help you avoid any potential errors related to the copying and pasting of code.

Download the example code files

You can download the example code files for this book from GitHub at `https://github.com/PacktPublishing/PHP-Web-Development-with-Laminas`. If there's an update to the code, it will be updated in the GitHub repository.

We also have other code bundles from our rich catalog of books and videos available at `https://github.com/PacktPublishing/`. Check them out!

Download the color images

We also provide a PDF file that has color images of the screenshots and diagrams used in this book. You can download it here: `https://packt.link/qw7ug`

Conventions used

There are a number of text conventions used throughout this book.

`Code in text`: Indicates code words in text, database table names, folder names, filenames, file extensions, pathnames, dummy URLs, user input, and Twitter handles. Here is an example: "The `.phtml` files have a combination of HTML and PHP code and the `.js` files contain exclusively JavaScript code."

A block of code is set as follows:

```
public function getArrayCopy()
{
    return get_object_vars($this);
}
```

When we wish to draw your attention to a particular part of a code block, the relevant lines or items are set in bold:

```
public static function encrypt(String $text)
{
    $text = strrev(hash('sha256', $text));
    return hash('md5', $text);
}
```

Any command-line input or output is written as follows:

```
Failed asserting response code "302", actual status code is
"404"
```

Bold: Indicates a new term, an important word, or words that you see onscreen. For instance, words in menus or dialog boxes appear in **bold**. Here is an example: "Save the file and click on the **Update dependencies** button."

> **Tips or important notes**
> Appear like this.

Get in touch

Feedback from our readers is always welcome.

General feedback: If you have questions about any aspect of this book, email us at customercare@packtpub.com and mention the book title in the subject of your message.

Errata: Although we have taken every care to ensure the accuracy of our content, mistakes do happen. If you have found a mistake in this book, we would be grateful if you would report this to us. Please visit www.packtpub.com/support/errata and fill in the form.

Piracy: If you come across any illegal copies of our works in any form on the internet, we would be grateful if you would provide us with the location address or website name. Please contact us at copyright@packt.com with a link to the material.

If you are interested in becoming an author: If there is a topic that you have expertise in and you are interested in either writing or contributing to a book, please visit authors.packtpub.com.

Share Your Thoughts

Once you've read *PHP Web Development with Laminas*, we'd love to hear your thoughts! Scan the QR code below to go straight to the Amazon review page for this book and share your feedback.

https://packt.link/r/1803245360

Your review is important to us and the tech community and will help us make sure we're delivering excellent quality content.

Download a free PDF copy of this book

Thanks for purchasing this book!

Do you like to read on the go but are unable to carry your print books everywhere? Is your eBook purchase not compatible with the device of your choice?

Don't worry, now with every Packt book you get a DRM-free PDF version of that book at no cost.

Read anywhere, any place, on any device. Search, copy, and paste code from your favorite technical books directly into your application.

The perks don't stop there, you can get exclusive access to discounts, newsletters, and great free content in your inbox daily

Follow these simple steps to get the benefits:

1. Scan the QR code or visit the link below

https://packt.link/free-ebook/9781803245362

2. Submit your proof of purchase
3. That's it! We'll send your free PDF and other benefits to your email directly

Part 1:
Technical Background

In this part of the book, you will obtain the necessary knowledge to prepare a free and open source development environment for object-oriented PHP web applications and an understanding of the basic concepts required to work with Laminas.

This section comprises the following chapters:

- *Chapter 1, Introducing Laminas for PHP Applications*
- *Chapter 2, Setting Up the Environment for Our E-Commerce Application*
- *Chapter 3, Using Laminas as a Library with Test-Driven Development*
- *Chapter 4, From Object-Relational Mapping to MVC Containers*

1

Introducing Laminas for PHP Applications

In April 2019, the Linux Foundation announced that it was launching a new project: **Laminas**. This was actually the new name for a project that had existed since 2006, **Zend Framework**. From that moment on, the PHP framework, which already had 400 million installs, would have a vendor-neutral home. But Zend Technologies has not abandoned the product that bears its name for several years. Zend continues to provide corporate support, which includes mission-critical support, enterprise adapter extensions, long-term support, and on-demand guidance from PHP experts.

This chapter presents the evolution of the architecture of Zend Framework until its change to Laminas and discusses the lessons learned. In this chapter, you will gain an understanding of why the framework team took some decisions instead of others during its evolution. These subjects are the motivation for a developer to use a framework, and Laminas is presented as one option for PHP developers.

In this chapter, we'll be covering the following topics:

- Why do we need web frameworks?
- From Zend Framework to Laminas
- The technical and social infrastructure of the Laminas community

Why do we need web frameworks?

Some people may ask, *Why do I need a framework?* Such people may also ask, *Why do I need to use classes?* After all, PHP is simple and it no needs classes. With a few lines of code, you can produce a page with processed data from a database. Rasmus Lerdorf, the PHP creator, always claimed in his talks that PHP is fast and it requires neither classes nor frameworks to work. In 2014, Mariano Iglesias gave a lecture at PHP Conference Brazil, in which he defended a future without frameworks. Well, both Rasmus and Mariano are right, depending on the software complexity we are talking about. We cannot compare a personal page with just a few statistics to an enterprise system. When we talk about **object-oriented programming** (OOP) and frameworks, we are mainly talking about complexity control.

Martin Fowler states that one of the meanings that most people agree on for the term **architecture** is a set of decisions that are hard to change. For a software project, these decisions are about the desirable characteristics of a piece of software. When you think about what is desirable for a piece of software, it is important to remember that the delivery of the first stable version of the software is only one milestone of its lifetime.

Continuous integration and continuous delivery

You must have heard of the terms **continuous integration** and **continuous delivery**. These approaches exist because change is continuous. Software can change for several reasons: new features might be needed because customers require them or because competitors already have them, bugs have to be fixed, computer platforms evolve, audiences grow, and legislation changes. So, customers, competitors, environments, and governments will force changes in software. We can have many doubts about the future, but one thing we can be sure of is that there will be changes.

> **Note**
>
> Jean-Marcel Belmont defines continuous integration as follows:
>
> "[...] *a software engineering task where source code is both merged and tested on a mainline trunk.* [...] *Continuous integration is continuous because a developer can be continuously integrating software components while developing software.*" You can read more about this in *Hands-On Continuous Integration and Delivery*, Jean-Marcel Belmont, Packt Publishing.
>
> Jez Humble and David Farley, in their book *Continuous Delivery*, Addison-Wesley, define this term as the release of new features of software *as frequently as possible* to reduce the risk that delivery carries.

A piece of software is a growing organism. Frederick Brooks, in his book *The Mythical Man-Month*, *Addison-Wesley*, states that there are four levels of complexity for software:

- A **program** whose algorithm can be stored in only one human brain; in other words, only one person can understand it. It is a very specific program with a very limited task.

  This kind of software serves specific needs and it is something that can be explained in a short conversation, although only the creator would be able to run and maintain it.

- A **programming product**: A program that can be run and maintained by anybody with minimal documentation and that is readable by human beings.

- A **programming system**: A set of programs where one program is combined with other programs to perform more complex tasks.

- **A programming systems product**: Software that is nine times more complex and expensive than a single program whose logic fits in the head of a single programmer.

  This software requires a big team to maintain it. It cannot fit in a single person's head. It is not easy to explain how it works or how it is organized. But, according to Brooks, it is the aim of most software development efforts.

> **Note**
>
> *The Mythical Man-Month* is a classic text for software engineering. Although it was written before the rise of agile software development, its analysis of the complexity of producing software and the difficulty in parallelizing tasks remains current.

Complex software under continuous change requires some characteristics to be maintained for a long time. We are talking about software made to last and not to throw away after a week. Some software is ephemeral, so you do not have to worry so much about its structure. Automatic code generation tools are great for generating code that will soon be discarded. But software that will evolve requires a deep understanding of how software works. In fact, software made to last requires some key characteristics: minimal complexity, loose coupling, extensibility, and reusability. Each one of these desirable characteristics is related to one critical characteristic: ease of maintenance.

It seems contradictory to say that a complex system must have minimal complexity. In fact, when we say that a *programming systems product* is complex, we are talking about a system that has many parts and that is big. Minimal complexity refers to the internal organization of these parts and how they are connected. It is not expected that every member of the development team knows everything about the entire system at all times, but it is desirable that any member of the team can easily understand a part of the system so that they can change it when it is necessary.

Erich Gamma, when he was interviewed by Bill Venners, stated that current software environments are too complex to create applications without the reuse of existing code. Gamma presumes that OOP is the standard paradigm of development because it offers the extensibility that is essential to reuse. Gamma states that there are three levels of reuse in OOP: reuse of classes, reuse of patterns, and frameworks. He emphasizes that frameworks are the highest level of reuse because they identify the key abstractions for solving problems and offer implementations that contain the knowledge and experience of several developers.

> **Note**
>
> You can read the complete interview with Erich Gamma at `https://www.artima.com/articles/erich-gamma-on-flexibility-and-reuse`.

The usefulness of a framework

When you use a framework to develop a piece of software, you are not only reusing classes like you do when you use libraries. Class libraries are like outsourcing in software development. You use class libraries because you want to focus on specific business issues, not on general issues that someone has already solved. But you call the classes of a library. A framework works under Hollywood's Law: *Don't call us, we'll call you.* The framework classes call your classes. It takes care of the general behavior of the software and lets you define the custom behavior. When you use a framework, it is like your software is a franchisee. Your software shares the same infrastructure that other software has. When a franchisor solves a problem or improves some feature of the business, it solves for all franchisees. When a framework offers a solution for a problem, it solves a common problem for all applications that use it. As the Dungeon Master says, *The fate of one is shared by all.*

We use a framework to make software easy to maintain. So, a framework has to help us to create code with minimal complexity and to reuse existing code with loosely coupled components. Once a framework implements design and architectural patterns, it provides a solution to generic issues, and so, in its evolution, it focuses on the improvement of generic solutions. Erich Gamma states that a framework provides default functionalities at a high level. The focus on the default issues of a piece of software allows that framework to be in control of an application, like a manager or an advisor.

When we see a great software developer like Mariano Iglesias speaking against frameworks, he is actually speaking against something else: **frameworkitis**. As Bill Venners had it in his conversation with Erich Gamma, this term means "*the disease of wanting to make frameworks out of everything.*" Erich Gamma says that some software named as frameworks actually wants to do too much for developers. The problem is that when they try to do it, they do so in a way that not all developers want and – even worse – in a way that developers cannot change.

A real framework must say, "*Do not worry, I control your application and will take care of the generic issues for you.*" A framework with frameworkitis, in turn, says, "*Do not worry, I generate code and do everything for you in the best way and you do not need to change my implementation – in fact, you cannot do so.*" Well, this behavior goes in the opposite direction of a constantly changing scenario. Frameworkitis takes developers from love to hate. First, they love a tool that seems to do everything for them, and then they hate the tool to which they are tied and that they cannot change. Thank goodness that Laminas is not like that.

From Zend Framework to Laminas

In the opening of the movie *X-Men* (2000), Professor Xavier says that mutation is the key to our evolution. He adds that mutation is how we have evolved from single-celled organisms into the dominant species on the planet. Software is like a living organism. It needs to evolve in order to survive. Sometimes it needs to change a lot and it seems like a mutation. The need for change requires that software allows changes. A hard-to-change piece of software is a candidate for extinction.

Evolution for a piece of software might look like replacing one component with another. A replacement may be necessary for the maintenance or improvement of scalability, performance, or security. But if it is hard or almost impossible to replace a component of a piece of software, the improvements will have to be made in a difficult way, usually by rewriting the existing code.

Patterns help with the replacement of components as components share the same patterns. If each development team of a company creates software using different patterns for components, it will be difficult to share components, prohibiting reuse, which is a desirable characteristic for software made to last.

My introduction to Zend Framework

In 2005, I joined a large IT company with several software development units. This company developed government software and had thousands of projects in various programming languages. The problem was that each team used different standards for their projects, so component reuse was virtually non-existent. In addition, there was a history of decisions made in the past by tools with low initial coding that in the long run precipitated technology lock-in. Several teams used frameworks, but each one of them used different frameworks for the same language. So, the difficulty in sharing components remained.

In 2007, I was working for a division that was prospecting for new technologies and my team had to choose a framework to standardize PHP application development in the company. By coincidence, that year a seminar on PHP frameworks took place in the city of São Paulo. We went to this event and attended three talks, each about a framework: CakePHP, Symfony, and Zend. I already knew CakePHP from the first PHP Conference Brazil and several colleagues from my company already used Symfony. Zend Framework, however, was a complete novelty for me. This should not be a surprise because the first stable version of Zend Framework was released that year. But it had been in development since 2005.

The audience reaction to the CakePHP and Symfony lectures was similar. When the speakers executed the commands in the terminal that generated code, the audience said, *Wow*! But in Zend's talk, when they saw that there was no code generator, they were disheartened. The speaker listened to the complaints and replied that although Zend did not have code generation tools, it was quite flexible, not imposing too much. Helpfully, one of the spectators noted that the framework architecture seemed to be concerned with maintaining the application and not just creating it.

On the way out of the event, I heard several people saying that they would use CakePHP or Symfony, emphasizing the code generation tools. These people probably did not have in-depth experience in software development, or rather in maintaining existing software. There are other tools that help with code generation, such as integrated development environments. Frameworks are reusable and are tools for reusing code, not new code creation tools.

One of the speakers was selling a book with an analysis of five frameworks, including the three presented at that event. I bought it, read it, and conducted tests for those frameworks, implementing the same application with each one of them. Among the five frameworks analyzed, the one that had the best characteristics of minimal complexity, loose coupling, extensibility, and reusability was Zend Framework. It was largely configurable and did not prevent the use of components from other frameworks. At the time, the **PHP Framework Interoperability Group** (**PHP-FIG**) did not yet exist, but we needed tools that offered interoperability.

> **PHP Framework Interoperability Group**
>
> In 2009, at PHP Conference Brazil, some PHP speakers gave a lecture called *PHP Framewarks*. It was a joking reference to a "war of frameworks." Each one of the speakers defended their favorite framework and attacked the others. Maybe there was not exactly a war of frameworks, but the fact was that once you adopted a framework, it was difficult to use the components of others because each one followed its own patterns. This scenario began to change in the same year, 2009, when a group of PHP experts met in Chicago, USA, for establishing specifications of components for the PHP community. This group was first called PHP Standards Group. The first specification was about an autoloading standard – what helps the emergence of Composer. In 2011, the group changed its name to **PHP Framework Interoperability Group** (**PHP-FIG**). Since then, members of awesome PHP projects have met to discuss and write specifications called **PHP Standard Recommendations** (**PSRs**). Several frameworks and other PHP applications have adopted the specifications of PHP-FIG.

Zend Framework – the reference architecture for PHP applications

Zend Framework allowed us to create a reference architecture for PHP applications that helped applications to evolve. An application can use internal components, which are within its structure and serve only it, or make use of reusable components, which are shared by several applications. A component can perform a task that extrapolates the responsibilities of the reference architecture pattern implementations or even replace the implementation of a layer (but using the same interface).

To solve a given problem, the software architect must first look for an existing component that solves it. If it exists, but does not meet all the requirements, the possibility of extending it should be analyzed. Only as a last resort should a new component be created. In this case, the component is born as an internal component to meet a specific application. But the component architecture must be treated as generic so that it can be later published and improved so that it becomes a reusable component. Zend Framework had a structure concerned with these issues.

The first stable release of Zend Framework was released before PHP 5.3, so ZF1 did not use namespaces. There was no Composer yet, so ZF1 had its own autoload implementation. But the one aspect that Zend Framework had present in its structure was, as I mentioned, the possibility to change. In fact, the unstable versions already brought configurable code like this:

```
Zend_Controller_Front::run('/path/to/controllers');
```

Instead of determining the directory for the controller classes, Zend Framework lets the developer choose. There was a proposed project structure, and today there is a **Model-View-Controller (MVC)** application skeleton available to start building an application, but the possibility to change the address of the project elements remains.

Another interesting feature of Zend Framework is recognizing that PHP applications already exist and you need to evolve what already exists. There will not always be demand for new projects, but what is in production needs to be maintained. Thus, Zend Framework allows from the beginning the use of components decoupled from the MVC implementation. Being able to refactor an application using Zend Framework components meant being able to outsource tasks that weren't the focus of the application's business without having to change everything. Change is necessary, but if it can be done slowly, gradually, and safely, it is better.

Zend Framework evolved gradually, keeping pace with the evolution of the PHP programming language and other software development technologies while incorporating design patterns and programming paradigms. In 2008, version 1.5 of Zend Framework incorporated the **Two-Step View** pattern with the `Zend_Layout` component and delivered a component to create dynamic forms, `Zend_Form`. In the same year, version 1.6 extended PHPUnit through the `Zend_Test` component. In 2009, Zend Framework gained a code generation tool, `Zend_Tool`, and a component for inversion of control, `Zend_Application`. IDEs such as Eclipse and Netbeans soon offered integration with `Zend_Tool`.

Even in its first major version, Zend Framework started to offer, even optionally, the possibility of creating modules. Modules, in Zend Framework terminology, are self-contained units that can be reused. Module architecture is well suited to complex systems that must be divided into several subsystems. In a world of microservices, modules allow you to create an initially monolithic application that can gradually be converted into a distributed system.

You might ask, *Why don't we just create an application using microservices?* The answer is another question: *Is that really necessary?* Today, we have elastic resources with cloud environments, but that is not a reason to allocate resources that we do not need yet. We do not need to change in advance but be prepared for change. The cake has an adequate amount of time to bake.

In version 2.0, Zend Framework evolved the module architecture and improved the consistency and performance of the components. The code was refactored to use namespaces and two new components were incorporated: `EventManager`, a combination of several strategies for event-driven programming used by the MVC implementation, and `ServiceManager`, a container for dependency injection. Decoupling between components has increased significantly since version 2.0.

Version 2.5 brought radical changes to Zend Framework. Each component was given its own version control repository and had an independent lifecycle. In addition, the autoloading of classes and the installation of components is now done by Composer. Until version 2.4, Zend Framework was a *download everything but just use what you want* framework. As of version 2.5, it became a *download and use only what you want* framework.

Version 3.0 of Zend Framework updated the minimum dependency on PHP to version 7, taking advantage of improvements implemented in the language. In addition, Zend Framework 3.0 brought a special microframework for creating middleware, **Expressive**. This microframework was implemented in strong compliance with PSRs from PHP-FIG, with complete independence from the components to be used.

And that was the end of Zend Framework. Yes, that was the end of it, because as we said at the beginning of the chapter, it is now called Laminas. *Lamina* means *blade*. The reason for this name is that the framework allows the creation of applications in layers, but thin layers, that is, with minimum necessary layer size. As Professor X said, evolution is slow. But sometimes evolution leaps forward. The change from Zend Framework to Laminas required a refactoring of all class names. In addition, the technical and social infrastructure of the community also changes. But this leap was good, because a framework with so many years of experience is more open now. And you are going to learn it here!

The differences between Zend Framework and Laminas

There are some differences between Zend Framework and Laminas as regards the community ecosystem. Zend Framework, as an open source project, accepted code contributions. Laminas, in turn, can also receive donations because it is under a non-profit organization. Zend Framework had forums and a Slack channel for the community to have discussions and make contributions. Laminas, in addition, has monthly open-to-the-public meetings and keeps a record of **Technical Steering Committee** votes (**TSC** is a group of members with technical knowledge that governs the project). In other words, Laminas is more open to community participation and transparent than Zend Framework. This is good! You can - *if you wish* – participate actively in the development of a product that helps you to create object-oriented web applications with several design patterns built in.

Steve McConnell states that the choice of programming language is a key construction decision for a software project. The PHP community, in turn, states that PHP is especially suited for web development. But the programming language is only one of the aspects you need to worry about when creating and maintaining web application projects. That's why we are talking about Laminas.

The technical and social infrastructure of the Laminas community

We will talk about many features of Laminas, but we will not cover everything, of course. The main reference source for Laminas is the official documentation, available at `https://docs.laminas.dev`.

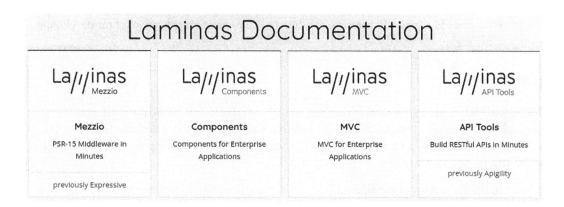

Figure 1.1 – Documentation home page for Laminas

The **Components** and **MVC** options on the documentation home page are directly related to the content of this book. Laminas, however, also includes the microframework **Mezzio** – formerly known as Expressive – and **API Tools**, an API generator built in with Laminas components.

The source code of all components is available for download and contribution at `https://github.com/laminas`. You will find more than 100 repositories with mature components. On this page, you can open issues for fixing bugs or requesting new features – remembering that for features, the level of response you get will depend on the interest of the community. If you have an idea for improving an existing component or even for creating a new component, be encouraged to send an implementation and not only a specification. Contributing to a free and open source project is an awesome way to learn more about software development and can help you to become known by your peers, who can then become your advisors. The guidelines for contributing to Laminas are available at `https://github.com/laminas/.github/blob/main/CONTRIBUTING.md`.

Roger Pressman defines software as a combination of computer programs and documentation about those programs. People are also part of a piece of software's documentation, because they often hold tacit knowledge that has not yet been converted into explicit knowledge. When documentation is not enough, we may need help from people, and a recorded conversation can become documentation. You can chat with people involved or interested in Laminas in the official discussion forums, available at `https://discourse.laminas.dev`.

You can report potential security vulnerabilities by sending a message to `security@getlaminas.org`. For information about the security policy of Laminas, read `https://getlaminas.org/security`.

You can get support from the community by opening issues and posting your questions on the forums; you can even solve issues yourself by extending components. But if you want *on-demand consultative support* or *long-term support*, you can think about the enterprise support of Zend Technologies. For more information about this service, read `https://www.zend.com/products/laminas-enterprise-support`.

After this overview of Laminas, in the next chapter, we will prepare an environment for developing with this framework.

Summary

Software is like a living organism: it is born, it grows, and someday it dies. We do not want our software to die early, of course. For this, we need to provide our software with characteristics that allow it to survive for a long time. In this chapter, we spoke about frameworks that help us to create reusable and easy-to-maintain software.

We touched on the history of Laminas, a free and open source framework for building PHP web applications. We presented its evolution and its proposal as a framework for evolving applications.

In the next chapter, we will prepare our environment to develop PHP web applications with Laminas.

2
Setting Up the Environment for Our E-Commerce Application

To build a house, we need various tools and materials. The more sophisticated the tools, the easier it will be to erect the house, and the better the materials, the longer the house will last. The same is true with software. Appropriate tools can help increase our productivity, and suitable components can make our software easier to maintain. In this chapter, we will set up the development environment for the example application that we will be building throughout this book: a virtual store that allows customers to be registered and buy products. The tools chosen are all free and **open source software (OSS)** with distributions for the most used operating systems in the world.

In this chapter, we'll be covering the following topics:

- Installing a generic **PHP: Hypertext Preprocessor (PHP)** environment, with a web server, PHP interpreter, and **relational database management system (RDBMS)**
- Two different ways to install and use Laminas
- Installing an **integrated development environment (IDE)**—the Eclipse platform
- Integrating Laminas-related commands into the Eclipse toolbar
- The data models we will use in this book for our exercises

Specifically, the chapter will comprise the following sections:

- Installing Laminas
- Configuring the IDE—Eclipse for PHP developers
- Integrating Laminas and Eclipse
- Managing MySQL

Technical requirements

In this book, we use Laminas MVC 3.3.4, which requires at least PHP 7.4. However, support for PHP 7.4 ended in 2022, so it is recommended that you use at least PHP 8.0, but be encouraged to use PHP 8.1. You can get deadlines for PHP support at `https://www.php.net/supported-versions.php`.

For creating and running the code of this book, you will need at least a PHP interpreter, a web server, and an instance of **MySQL/MariaDB** (although you can use every database supported by PHP if you adapt the database script). There are several ways to install each software, and there are some packages that reunite them. **XAMPP** is a package that bundles the **Apache HTTP Server**, PHP, and MySQL/MariaDB. You can download XAMPP for Windows, Linux, or macOS from `https://www.apachefriends.org/download.html`. Since it can be used with most operating systems of the world, we adopt it as our default PHP environment. Moreover, XAMPP provides a control panel that allows you to easily start/stop Apache and MySQL. In the following screenshot, we show a sample of the XAMPP control panel for Linux:

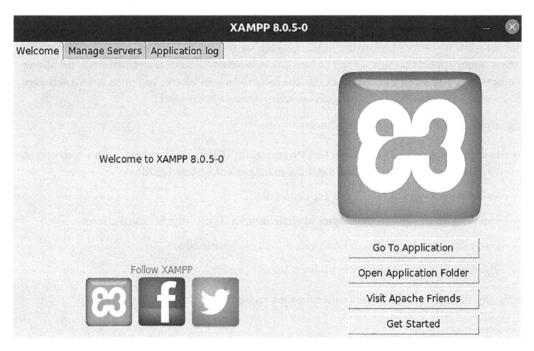

Figure 2.1 – Window of XAMPP control panel

The home page of *Apache Friends*, the host project of XAMPP, has a short and nice video explaining the installation process: `https://www.apachefriends.org/index.html`.

XAMPP also brings a **phpMyAdmin** instance for managing MySQL databases. Once you have started Apache and MySQL through the XAMPP control panel, you can access phpMyAdmin at `https://localhost/phpmyadmin`.

We will use the command line sometimes. For avoiding the use of the complete path of the PHP interpreter in the terminal, you must add the directory containing the PHP interpreter to your `PATH` environment variable. PHP is inside the following path:

- `c:\xampp\php` for Windows
- `/opt/lampp` for Linux
- `/Applications/XAMPP/bin` for OS X

After installing your PHP environment, you must install **Composer**, a dependency manager for PHP. You can download it from `https://getcomposer.org/download`. Put the `composer.phar` file in the same directory as the PHP interpreter. For Linux and OS X users, rename the file `composer` only and grant it permission to run. For Windows users, follow the instructions at `https://getcomposer.org/doc/00-intro.md#installation-windows`.

From now on, our workspace will be the `htdocs` folder of the XAMPP installation. We will create all our examples in this folder. Remember: when I am talking about **workspace** in this book, I am talking about `htdocs`. Make sure the `htdocs` folder has read and write permissions so that you can run commands into it with your user and open files with the IDE.

> **Note**
> If you already have a web server, PHP, Composer, and MySQL installed, you can use them, but in this book, we presume the use of XAMPP, so references about directory structure and programs are related to XAMPP.

Installing Laminas

You can install Laminas through each of the following approaches:

- Installing an isolated component of Laminas for use with an existing PHP application (or a small application without MVC)
- Installing the skeleton application for a new application with an architecture based on MVC

Installing Laminas through an isolated component of Laminas

Let's take an example of installing the **object-relational mapping** (ORM) and database abstraction component, `laminas-db`. Inside the `htdocs` folder, run the following command:

```
composer require laminas/laminas-db
```

This command will create a folder named `vendor` and download the `laminas-db` component and its dependencies. For loading the classes of the component, you need to import the `vendor/autoload.php` file. For a while, you only need to check whether Composer is working. We will build basic examples with isolated components later, so do not worry. You can remove `laminas-db` after downloading it successfully. To only remove the component, you run the following command:

```
composer remove laminas/laminas-db
```

This command will remove the `laminas-db` component within the `vendor` folder, but it will keep the `vendor` folder. This folder (`vendor`) in the root of `htdocs` will not be used, so you can delete it. We will create a specific folder for our projects.

Installing Laminas through an MVC skeleton application

For starting an MVC web project with Laminas, you have three alternatives, as outlined here:

1. Clone the skeleton via the `https://github.com/laminas/laminas-mvc-skeleton` Git repository.

    You have to delete the control version after cloning. You can do so by accessing the `laminas-mvc-skeleton` folder and removing the `.git` folder inside it. The `.git` folder is part of a Git repository structure. It doesn't appear in the page of GitHub but is created when you clone the repository. After that, you can rename the `laminas-mvc-skeleton` folder with the name of your project.

2. Download the skeleton via the following link: `https://github.com/laminas/laminas-mvc-skeleton/archive/refs/heads/1.3.x.zip`.

    Decompress the file and rename the folder `laminas-mvc-skeleton`.

3. Install the skeleton with Composer.

We will use the third alternative. For this, run the following command:

```
composer create-project -sdev laminas/laminas-mvc-skeleton
myfirstlaminas
```

This command output will start with the following message:

```
Creating a "laminas/laminas-mvc-skeleton" project at "./
myfirstlaminas"
```

After downloading the files from the repository, the project creator will ask you the following:

```
Do you want a minimal install (no optional packages)? Y/n
```

Now, you must answer Y because it is only a preview for the next exercises.

After that, you will have a folder named myfirstlaminas. We will start a web server pointing to the public subfolder as the web root. For this, run the following command:

```
php -S localhost:8000 -t myfirstlaminas/public/
```

Of course, if you are running on Windows, the directory separator is \.

You may receive this error message:

```
Failed to listen on localhost:8080 (reason: Address already in
use)
```

If so, this is because port 8000 is used by another process. No problem—you can use 8008 or another available port.

Considering that you could start the web server successfully at port 8000, if you type the address http://localhost:8000, you will be able to see the home page shown in the following screenshot:

Figure 2.2 – Home page of MVC Skeleton Application

Try requesting a non-existing page, such as `http://localhost:8000/otherpage`. In this case, you will see a page like the one shown here:

A 404 error occurred
Page not found.

The requested URL could not be matched by routing.

No Exception available

© 2021 Laminas Project a Series of LF Projects, LLC.

Figure 2.3 – Page not found error page

These pages show that Laminas MVC components are installed successfully. For our exercises, the built-in web server of PHP is enough because we are in a development environment.

But if you want to use a production web server, there are objective references about Apache and nginx configuration for Laminas, respectively, as follows:

- `https://docs.laminas.dev/tutorials/getting-started/skeleton-application/#using-the-apache-web-server`

- `https://docs.bitnami.com/installer/infrastructure/nginx/get-started/use-laminas`

We can say now we made a "Hello World" application with Laminas. The code of the framework is installed and it is producing visible results. We have the bricks for building a house, but carrying the bricks only with our hands can be difficult. Sometimes, it is good to use a pushcart. Other times, it can be better to use a truck. Let's use a truck for carrying many software bricks: an IDE.

Configuring the IDE – Eclipse for PHP developers

Laminas does not depend on any IDE to work. In fact, you can use Vi or **Visual Studio Code** (**VS Code**) to program in PHP; for the final result, there is no difference. However, the programming activity can be very complex and full of details, and it is easy to get lost searching for the cause of a bug. Specifically, it is very easy to deliver some results with PHP, but it is also very easy to lose control as the single program grows and becomes a complex application. We need tools that help us to control the complexity of our software systems and let us focus on implementing the use cases or stories for which our customers pay us.

In this book, we suggest the use of the **Eclipse** platform as our IDE for Laminas. Eclipse has the feature to be pluggable—you can add and remove tools according to your needs. Many companies and communities provide plugins for Eclipse, and the platform has a marketplace where you can find hundreds of components integrating Eclipse with other systems. You can adjust Eclipse for your reality instead of adjusting your work process for some tool.

For your convenience, there are already many distributions of Eclipse for specific purposes. Among them, there is one for PHP: **Eclipse PHP Development Tools** (**PDT**). You can download it from `https://www.eclipse.org/pdt/#download`. Eclipse requires the **Java Development Kit** (**JDK**). Install the JDK if you do not have it yet, after which you should download PDT. The installation is essentially a file uncompress. In the end, you will have an executable file called `eclipse.exe` (Windows) or only `eclipse` (Linux/OS X) inside the `eclipse` folder. When you run this file, Eclipse will open a window asking for the workspace. You must put the path to the `htdocs` folder of XAMPP, as seen in the following screenshot:

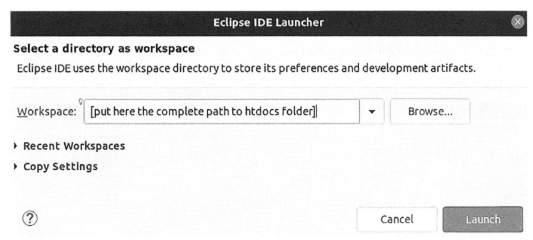

Figure 2.4 – Setting of Eclipse workspace

Once the workspace is defined, Eclipse will open your main window with the **Welcome** internal window, as we see in the following screenshot. You can close this window because we will not need it:

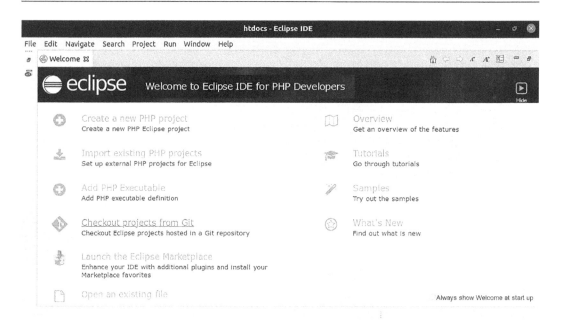

Figure 2.5 – Welcome view of Eclipse PDT

Do not worry about losing the links in the **Welcome** window because all the links inside it are available in the Eclipse menu. After closing the **Welcome** window, you will see a set of internal windows, as shown in the following screenshot:

Figure 2.6 – Welcome view of Eclipse PDT (continued)

On the left, you can see the **Project Explorer** window, where PHP projects are listed in a tree structure. On the right, you can see a window named **Outline**, where the structure of open files in the central window can be seen; this is a container for different text editors. At the bottom, you can see the **Problems** window, which shows error messages and warnings related to open files. These windows and others that are not relevant to our exercises compose the *PHP Perspective*.

> **Views and Perspective**
>
> There are two core concepts in an Eclipse interface: views and perspective. An internal window is called a *view*, while a set of views is called a *perspective*. You can open any view by clicking on **Window** | **Show View**. Similarly, you can open a perspective by going to **Window** | **Perspective** | **Open Perspective**.

Eclipse, of course, has a lot more features. Do not worry, because we will talk about each one of the features related to development with Laminas as we need them. Next, we will learn how we can integrate any program with Eclipse and why this is useful for working with Laminas.

Integrating Laminas and Eclipse

After integrating the command for creating Laminas projects, we need to configure our PHP environment in Eclipse.

You do this by following the next steps:

1. Go to **Window** | **Preferences** | **PHP**.
2. The first item to define is the PHP executable.
3. Select the **Installed PHPs** item on the left, as seen in the following screenshot, and click on **Add...** on the right:

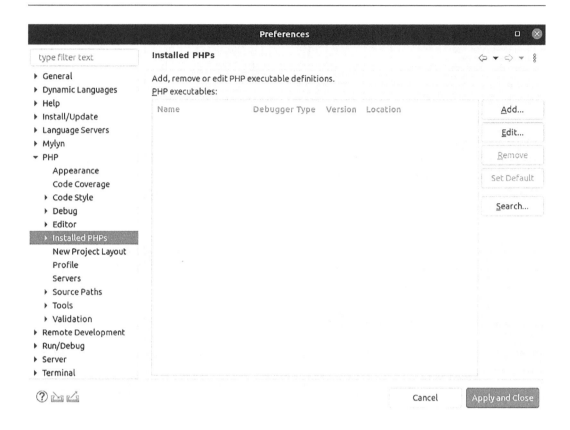

Figure 2.7 – Preferences for PHP environment

4. A window will open, as seen in the following screenshot. You must configure many PHP executables, so each one of them must have a different name. Here, we will work with only one called PHP:

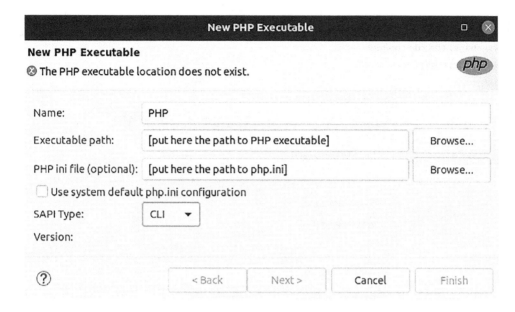

Figure 2.8 – Configuration of PHP executable

5. Fill the paths for the PHP executable and php.ini file.

6. Select the **SAPI Type** value and click on **Finish**.

Now, we will configure the call for the create-project command of Composer inside Eclipse. You do this by clicking on **Run | External Tools | External Tools Configurations...**, to reach the following screen:

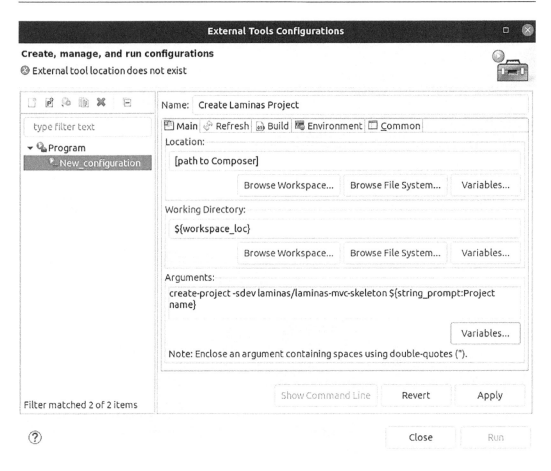

Figure 2.9 – Configuration for Composer create-project

Next, double-click on **Program** on the left of the screen shown in *Figure 2.9* to create a new program configuration. On the right, you will see a **Main** tab that you will fill out in the following manner:

- **Name**: `Create Laminas Project`

- **Location**: Here, you must type the complete path to the Composer executable

- **Working Directory**: `${workspace_loc}`

- **Arguments**: `create-project -sdev laminas/laminas-mvc-skeleton ${string_prompt:Project name}`

Expressions such as ${ . . . } are Eclipse variables. The ${workspace_loc} variable contains the complete path to the current Eclipse workspace. The ${string_prompt} variable, in turn, contains the text typed by a user in a dialog box. This dialog box opens when Eclipse runs the program. You can write a message to appear in this dialog box, putting text after the term string_prompt, preceded by a colon. So, for showing the Project Name message, you write ${string_prompt:Project name}. There are other variables, and you can see all of them by clicking on the **Variables** button.

The **Main** tab shows the working part of the program, but for the program to appear in the Eclipse toolbar, we have to select an option under the **Common** tab, as illustrated in the following screenshot:

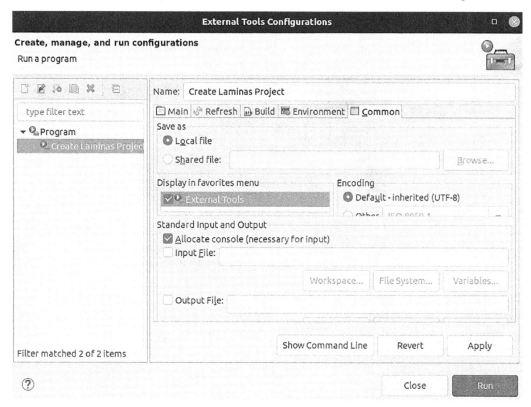

Figure 2.10 – Enabling the program as an item in the menu

As we can see in *Figure 2.10*, the **Common** tab has a frame named **Display in favorites menu**. You must check the **External Tools** option inside the frame for the program to be visible in the menu. After that, you can click on **Apply** to save the configuration.

Then, you can click on the **External Tools** button in the toolbar. This button's icon is a green play symbol with a red toolbox. As we can see in the following screenshot, the button will open a list with our Create Laminas Project program:

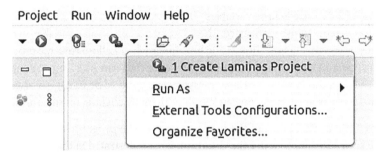

Figure 2.11 – Create Laminas Project command in the menu

Clicking on **Create Laminas Project** will open a dialog box, as shown in the following screenshot. Type mysecondlaminas as the project name and click on **OK**:

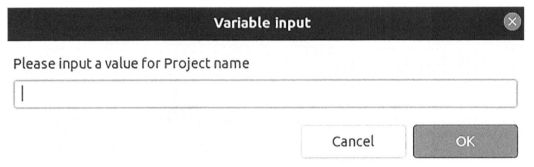

Figure 2.12 – Dialog box of Create Laminas Project command

A view named **Console** will open at the bottom and show you the same output you would see in the terminal if you were to run the command there. You know that is over when <terminated> appears at the top of the **Console** view, as we see in the following screenshot:

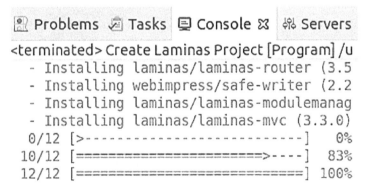

Figure 2.13 – Console view

To open the new project, you must access the **File | New | PHP Project** menu shown in the following screenshot. Fill the **Project name** field with mysecondlaminas and click on **Finish**:

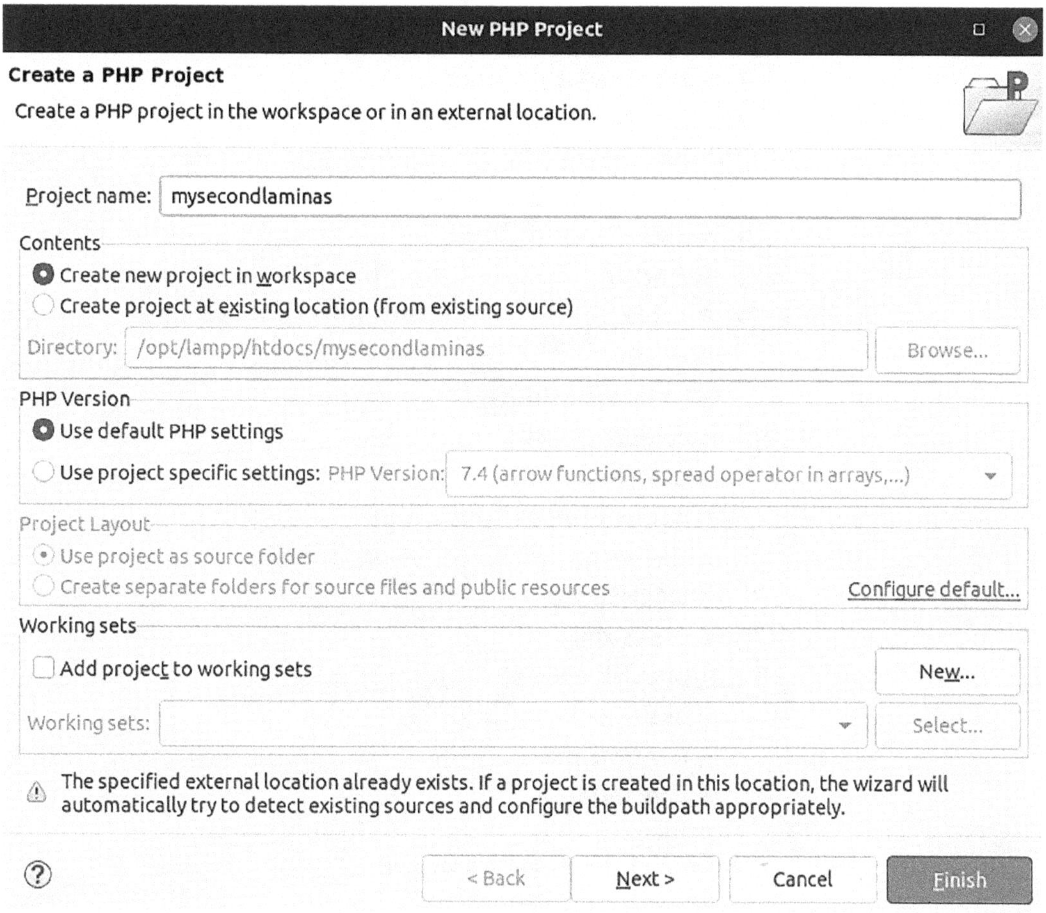

Figure 2.14 – New PHP Project screen

The mysecondlaminas project will now be open in the **Project Explorer** view. If you click on the project name, it will show its content, as we can see in the following screenshot:

Figure 2.15 – mysecondlaminas project

The project was not created by Eclipse but associated with Eclipse because it already existed. Now, Eclipse recognizes the mysecondlaminas folder as a PHP project and enables all features related to this programming language.

Now you know how to create a Laminas project with Eclipse, you can remove the mysecondlaminas project. For this, you must right-click on the project name and choose **Delete**. This will open the following dialog box:

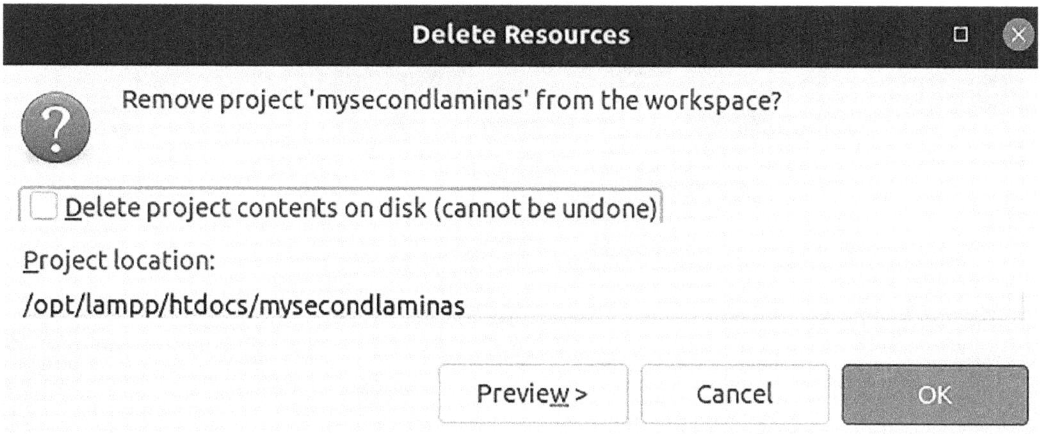

Figure 2.16 – Delete Resources dialog box

Calm down now. Wait. If you click on **OK**, you will unassociate the project as an Eclipse project. It will exist in the filesystem. To really remove the project and delete all files, you must check the **Delete project contents on disk (cannot be undone)** option. Yes, it is a scary message, but you can check it and remove the `mysecondlaminas` project because we will not need it anymore. Yes, this time you can click on **OK**—no fear.

If we need more integrations between Eclipse and other programs, we will make these as appropriate. For now, it is enough to know you are able to integrate programs into the Eclipse toolbar. Moreover, you also may be anxious to make something—I want to say, some code to run. Be calm—before starting to code, we need to define data models for the exercises. Let's go!

Managing MySQL

As we said before, XAMPP brings a phpMyAdmin instance for managing MySQL databases, accessible at `https://localhost/phpmyadmin`. By default in XAMPP, the MySQL superuser, `root`, has no password. So, when you access phpMyAdmin for the first time, you land on the phpMyAdmin home page, as shown in the following screenshot:

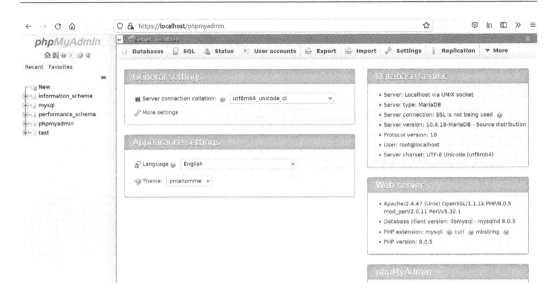

Figure 2.17 – Home page of phpMyAdmin

There are two frames on this page: on the left, you see a databases tree, and on the right, you see an environment overview.

As you can perceive, there are already some system databases, such as information_schema and performance_schema. There is also a node called **New**. Click on **New**, and you will see on the right the frame shown in the following screenshot:

Figure 2.18 – Home page of phpMyAdmin (continued)

You can use this frame to create databases that will be used in our exercises. It is enough to fill out the input text database name and click on **Create**. We need to create two databases: `school` and `whatstore`. Create them. You will be able to see them on the left. Do not worry about the tables because we will use scripts to create them. These scripts are available at `https://github.com/PacktPublishing/PHP-Web-Development-with-Laminas/tree/main/databases`.

The `school` database will serve for *starting the engine*. It has only two tables: `classes` and `students`. It is a cut of a school system, but there is a one-to-many relationship. The table structures with fields and data types can be seen in the following screenshot:

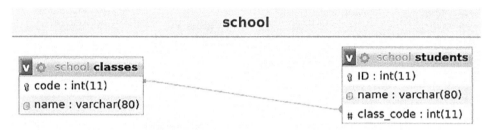

Figure 2.19 – Data model of school database

The `whatstore` database will serve our challenging exercise—the e-commerce application. It has 12 tables, divided into 2 groups.

One group is the business core, composed of the following tables:

- `customers`
- `discounts`
- `inventory`
- `order_items`
- `products`
- `purchase_orders`
- `transactions`

The other group is the management module, composed of the following tables:

- `employees`
- `employee_roles`
- `resources`
- `resources_role`
- `roles`

The table structures with fields and data types can be seen in the following screenshot:

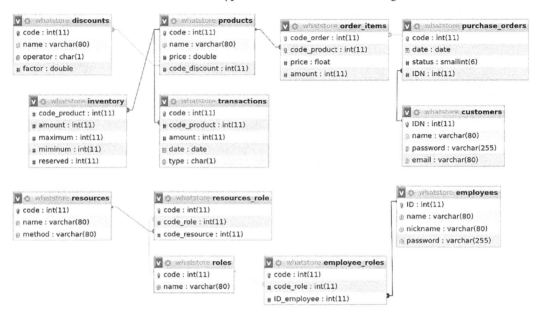

Figure 2.20 – Data model of whatstore database

The script for the school database is the school.sql file, and the script for the whatstore database is whatstore.sql. You can download them from the GitHub repository.

I told you what the **Uniform Resource Locator** (**URL**) is some paragraphs before—remember? To create tables from these scripts, we can use the importing page of phpMyAdmin. First, you select a database on the left-hand side, clicking on the name. For example, let's select the school database. After selecting this database, click on the **Import** tab on the right frame. Then, you will see a page like the one shown in the following screenshot. To import the script, click on the **Browse** button and search for the file in your filesystem. For the school database, the script is school.sql, right?

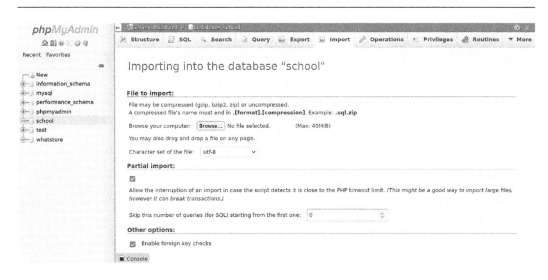

Figure 2.21 – Importing page of phpMyAdmin

After you have selected the script file, you must click on **Go**, at the bottom of the page on the right, as we can see in the following screenshot:

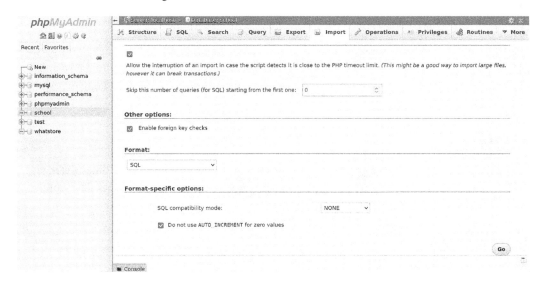

Figure 2.22 – Go button at the bottom of importing page

Wait for the processing of the script and you will see the following page, with the results of importing:

Figure 2.23 – End of a Structured Query Language (SQL) schema importing process

On the left, we will see new table nodes. For the `school` database, we will see the `classes` and `students` nodes. Now, do the same process for the `whatstore` database.

We have finally concluded the configuration of our development environment. We have all things we need to practice our exercises. I hope you have followed successfully the guidelines for the environment to avoid bad surprises in the next chapters. According to Benjamin Franklin, "*by failing to prepare you are preparing to fail.*"

Summary

In this chapter, we have installed a generic PHP environment, with a web server, PHP interpreter, and RDBMS. This environment is based on XAMPP, a package of free software and OSS for development with PHP, available for the most used operating systems.

We have learned two different ways to install and use Laminas: as isolated components to be used by PHP applications with any type of structure, whether MVC or not; and as a container for an MVC application, delegating the control of the application to the framework.

We also installed a free and open source IDE: Eclipse for PDT. We have shown how to integrate programs with the **command-line interface** (**CLI**) into the Eclipse toolbar, configuring an item to create a Laminas project within Eclipse.

Finally, we created the databases for this book's exercises—`school` and `whatstore`—with the help of phpMyAdmin, embedded into XAMPP.

In the next chapter, we will learn the foundations of programming with Laminas through some exercises with the `school` database.

3

Using Laminas as a Library with Test-Driven Development

The next two chapters will prepare you to use Laminas. The preparation consists of three stages to allow you to gradually adapt to writing code in Laminas. In this chapter, we will go through the first stage, which we will explain using a simple example based on a reduced use case, showing it is possible to only use one Laminas component in an existing PHP application. We will initially use the laminas-db component to abstract a database connection. In the next chapter, we will go through the other two stages.

You could hypothetically jump from the previous chapter to *Chapter 5, Creating the Virtual Store Project*, but it would be as though someone gave you a lightsaber without any training and told you to face Darth Vader in your first mission. Instead, let us keep calm and prepare ourselves properly for battle.

In this chapter, we'll be covering the following topics:

- Understanding the software requirements
- Developing our use case with TDD
- Generating code coverage reports

Technical requirements

To implement the example in this chapter, the following items are required:

- XAMPP 8 or higher
- Eclipse for PHP Developers 2021-12 or higher
- The school database created in the previous chapter

The installation or creation of all of these items is explained in *Chapter 2, Setting Up the Environment for Our E-Commerce Application*.

The example in this chapter was originally developed on a computer with 7.6 GB RAM, 1.0 TB disk space, and a CPU with 4 cores of 2.50 GHz.

The code generated for this chapter is fully available at `https://github.com/PacktPublishing/PHP-Web-Development-with-Laminas/tree/main/chapter03`.

Understanding the requirements for developing our software

We need the right requirements for developing the software – if we have the wrong requirements, we will produce the wrong software. It does not matter if we have a fantastic framework, a spectacular IDE, or an amazing continuous integration environment. If we have good tools, but the wrong requirements, we will only deliver the wrong software faster.

> **Very important note**
> A framework won't fix errors in your requirements.

Without further ado, let's get to our requirements. Imagine you are working on a system for school administration. More precisely, you have been charged with developing the subsystems for managing classes and students. As you will notice, there are several details in common between these groups of school elements. We are highlighting the similarities because this is convenient as we work with code reuse.

Imagine that a careful requirement determination was conducted and you have received the following use case for implementing the registration of classes first.

Use case – managing school classes

The common preconditions for the flows of this use case are the existence of a database named `school`, which has a table named `classes`, and a valid connection between the application and the database. The common exception is an SQL error. Here is the list of flows we will need for our use case:

- **Listing flow (basic)**

  - Description: A user requests a list of classes

- **Inserting flow (alternative)**

  - Description: A user inserts a new class
  - Postcondition: To find a record in the table with the input data

- **Updating flow (alternative)**

  - Description: A user updates an existing class

- Postcondition: To find the changes in the record

- **Deleting flow (alternative)**

  - Description: A user deletes an existing class

  - Postcondition: Not to find the record anymore

Now that you know about the flows we will need for our use case, let us start implementing a structure for this use case.

Implementing a structure for the use case

Now that we have our requirements, let's create a PHP project named `school1` using the following steps:

1. In Eclipse, access **File | New | PHP Project**.

2. In the **New PHP Project** window (see *Figure 3.1*), type `school1` and click **Finish**:

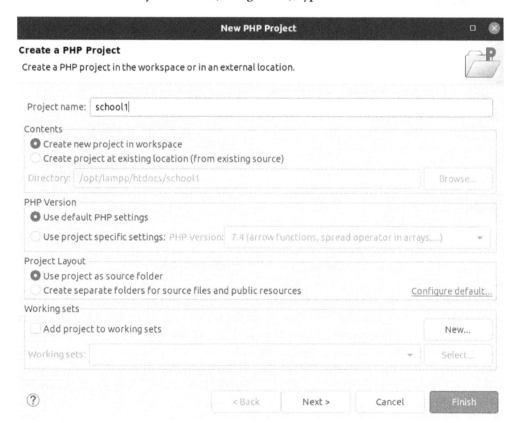

Figure 3.1 – Creating the school1 project

After that, you will see a node named **school1** in the **Project Explorer** view, as we can see in *Figure 3.2*:

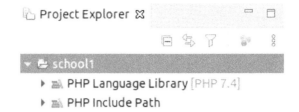

Figure 3.2 – The school1 project in the Project Explorer view

Did you think that something was wrong because the PHP version displayed on the right-hand side of the subnode, **PHP Language Library**, is lower than the installed version? It seems strange because you have configured the PHP interpreter before. However, if you access the **Window | Preferences** menu and select the **PHP | Installed PHPs** item, you will see the right version of PHP, as shown in *Figure 3.3*:

Figure 3.3 – The configured PHP interpreter in Eclipse

Well, there is nothing wrong. The version that appears in the **Project Explorer** view is the version of the syntax and token validation for PHP. This means the Eclipse code editor for PHP will validate the elements defined until that version. If you are upset because you want to use the latest features immediately, I ask you to be calm. You can use the latest features without any problems because only what the interpreter understands matters. For example, PHP 8 has named arguments for functions but PHP 7 does not. If you use named arguments in a PHP file, the Eclipse code editor PHP (prepared for PHP 7.4) will highlight the named argument as an error. However, if you are using a PHP 8 interpreter, the code will run successfully. In other words, ignore error messages about features exclusive to PHP 8. They are false alerts that will disappear in the next release of Eclipse.

More important than the warning about static code is to have tests for running code. Yes, we will create tests. For a while, we will write tests after writing some implementation – something not so good. Why is this not good? If you implement something before creating a test for it, you may neglect the tests. But if you create the test first, you ensure that something checks whether your code works. It is for this reason that we will create the tests before the complete implementation, in the next section, according to the guidelines of **test-driven development (TDD)**.

Please see the architecture draft constructed using a **Unified Modeling Language (UML) component diagram** here (in *Figure 3.4*) for the use case of registering the classes. This is the only use case we will implement in the school1 project:

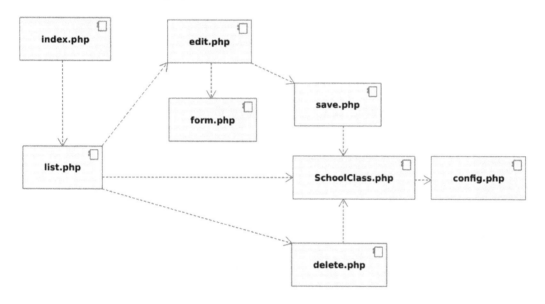

Figure 3.4 – A UML component diagram for the files of the school1 project

As you can see in *Figure 3.4*, we planned the registration of the school classes as a set of seven PHP files. The purposes of these files are as follows:

- index.php: Refers to the homepage, the starting point for the actions
- list.php: Shows the list of classes with hyperlinks for insert, update, and delete classes
- edit.php: Shows a form to insert or update a class
- form.php: Defines the form for edit.php
- save.php: Inserts or updates a class into a database
- delete.php: Removes a class from a database

- `SchoolClass.php`: A model for a school class that encapsulates the handling of the table classes

- `config.php`: Contains the configuration data for the connection to the database

The attributes and methods of the class named `SchoolClass` are presented in the class diagram in *Figure 3.5*:

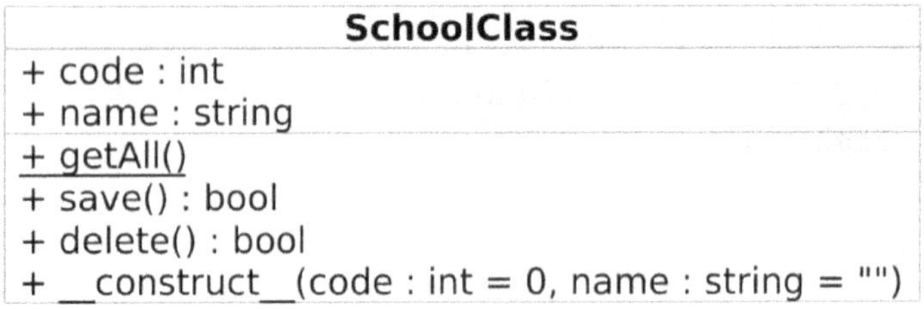

SchoolClass
+ code : int
+ name : string
+ getAll()
+ save() : bool
+ delete() : bool
+ __construct__(code : int = 0, name : string = "")

Figure 3.5 – A UML class diagram for SchoolClass

The `SchoolClass` class has attributes equal to the fields of the table classes. This class also has the following methods:

- `getAll`: Returns the data of the listing flow

- `save`: Executes the inserting and the updating flows

- `delete`: Executes the deleting flow

- `__construct__`: Initializes an instance with the data of a specific record

> **Note: Laminas does not require a fully object-oriented project**
>
> You don't have to use **object-oriented programming** (**OOP**) with PHP. You are free to only use structured programming, only OOP, or a combination of both paradigms. Laminas does not imprison you to OOP even though it is implemented under OOP. It does not impose the overall structure of your software project, so you can use classes according to your real needs.

Well, as we have minimal project documentation, let us implement our school class registration using the following steps:

1. In Eclipse, select the `school1` project, access the **File** | **New** | **PHP File** menu, and type `index.php` into the **File name** field.

2. The PHP file editor will open a view at the center of the Eclipse window with one line beginning with `<?php`.

3. Delete all the content in this file.

4. Then, press *CTRL* + the spacebar. You will see a list of templates, as we can see in *Figure 3.6*:

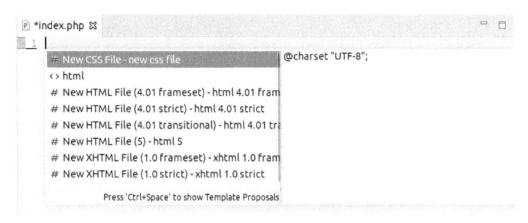

Figure 3.6 – Using a template for an HTML document

5. This feature allows us to fill a web file with repetitive content that you always have to write, such as document opening and closing tags.

6. Choose the **New HTML File (5) – html 5** item. Eclipse will fill the file with an HTML 5 basic structure. We will add our specific content according to the following source code:

index.php

```
<!DOCTYPE html>
<html>
    <head>
        <meta charset="UTF-8">
        <title>Classes Registration</title>
    </head>
<body>
  <h1>Classes Registration</h1>
  <p>
    <a href="edit.php">Add a new class</a>
  </p>
  <table>
    <thead>
      <tr>
        <th>code</th>
        <th>name</th>
      </tr>
    </thead>
```

```
    <tbody>
<?php
require 'list.php';
?>
    </tbody>
  </table>
</body>
</html>
```

7. As we have shown in the component diagram (see *Figure 3.4*), the index.php file depends on the list.php file. The first file keeps the static content while the second file generates the dynamic content. Since we have created the first file, we will now create the second file.

8. Select the school1 project, access the **File** | **New** | **PHP File** menu, and type list.php into the **File name** field. Input the following code into the file:

list.php

```
<?php
use School\SchoolClass;
foreach(SchoolClass::getAll() as $row):
?>
<tr>
<td><a href="edit.php?code=<?=$row['code']?>"><?=$row
    ['code']?></a></td>
<td><?=$row['name']?></td>
<td><a href="delete.php?code=<?$row
    ['code']?>">Remove</a></td>
</tr>
<?php
endforeach;
```

9. The list.php file contains a repeating loop that iterates over a collection provided by the SchoolClass class for listing the school classes. We are using an alternative syntax of the foreach structure, which uses endforeach instead of braces. This alternative syntax is useful when you have PHP code within HTML code because it is clearer. We have a few lines of code now, but as we have more and more lines, it can become confusing to understand the many different openings and closings of the PHP code blocks.

10. There must be warnings in the code editor because we have not created the class `SchoolClass` yet. As we can see in *Figure 3.7*, the code editor shows a warning on the left- and right-hand sides. On the left-hand side, the editor shows an icon of a lamp with a white X inside a red square. On the right-hand side, the editor shows red rectangles:

```php
list.php
1  <?php
2  use School\SchoolClass;
3
4  foreach(SchoolClass::getAll() as $row):
5  ?>
6  <tr>
7  <td><a href="edit.php?code=<?=$row['code']?>"><?=$row['code']?></a></td>
8  <td><?=$row['name']?></td>
9  <td><a href="delete.php?code=<?=$row['code']?>">Remove</a></td>
10 </tr>
11 <?php
12 endforeach;
```

Figure 3.7 – The error warnings on either side

11. The icons on the left-hand side allow us to try and fix the errors. When you click on any of them, Eclipse will show suggestions for fixing. The rectangles on the right-hand side highlight the errors in the line.

This concludes the basic structure for the use case of managing school classes. Next, we will create tests before implementing more business code.

Creating our first automated test

Let us now create our first automated test for checking whether the `SchoolClass::getAll` static method returns what we expect. But... what do we expect? A collection of arrays, each with the `code` and `name` keys.

Before creating the first test case, let's install **PHPUnit** into our project. We will use **Composer** for this; however, this time, we will use Composer inside Eclipse.

To install PHPUnit, follow these steps:

1. Right-click on the `school1` project and access the **Configure | Add Composer Support** menu, as shown in *Figure 3.8*.

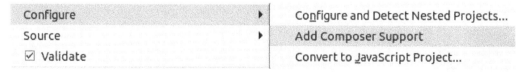

Figure 3.8 – Adding Composer Support

2. A file named `composer.json` will be created in the root of the project. Eclipse will open a view for editing the `composer.json` file:

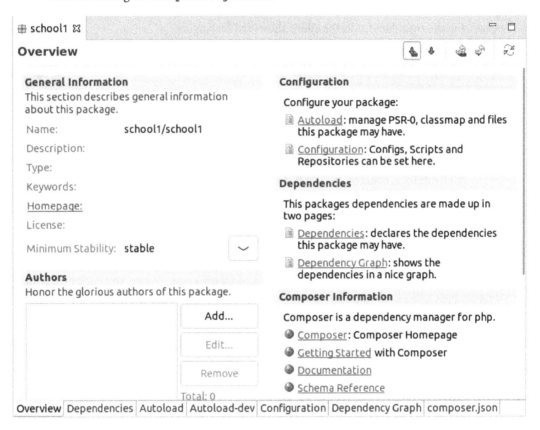

Figure 3.9 – The Composer view

3. The Composer view has several tabs at the bottom. One of them is **Dependencies**. This tab fills the **composer.json** section related to project dependencies.

4. Click on the **Dependencies** tab and the view will show two sections for including dependencies:

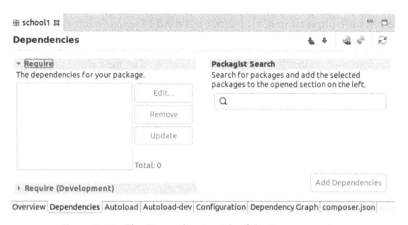

Figure 3.10 – The Dependencies tab of the Composer view

5.   We will add a development dependency, PHPUnit, a test framework for PHP applications.

6.   Click on the **Require (Development)** section at the bottom of the screen and this section will expand inside the view.

7.   On the right-hand side, there is a search box called **Packagist Search**. We can install any dependency available at the repository (`https://packagist.org`) here.

8.   Type `phpunit` into it and the plugin will list all the available packages whose names match `phpunit`. The search is performed using partial, not exact, matches (see *Figure 3.11*):

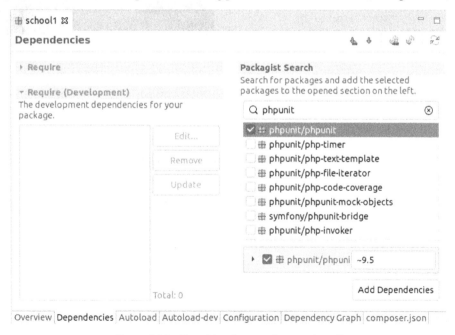

Figure 3.11 – Searching for a package to install

9. Select the **phpunit/phpunit** item and click on the **Add Dependencies** button. A package of PHPUnit in the latest available version will be inserted into the list on the left-hand side.

10. Click on the **composer.json** tab to view the code generated by the Eclipse plugin for Composer:

composer.json

```
{
    "name" : "school1/school1",
    "require-dev" : {
        "phpunit/phpunit" : "~9.5"
    }
}
```

11. Don't forget to hit *CTRL + S* for saving the changes. The plugin for Composer generates code but doesn't save it automatically.

12. At the top of Composer view, on the right, there are some buttons. The first button on the left is **Install dependencies**:

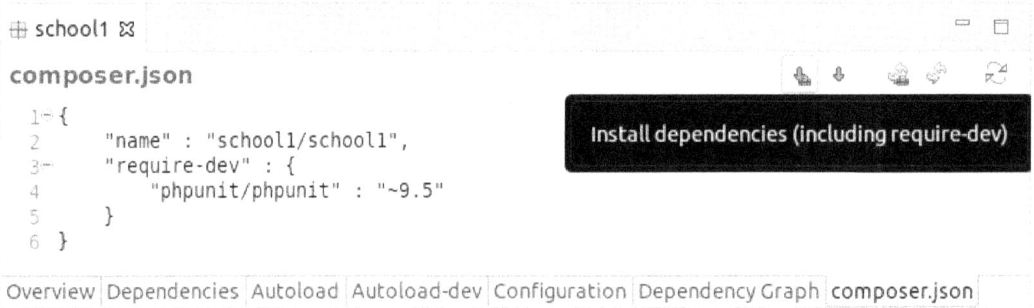

Figure 3.12 – Installing dependencies with Composer

13. Click on the **Install dependencies** button and Eclipse will run the `composer install` command.

14. You will see that Composer will create a folder named `vendor` and two files, `composer.lock` and `composer.phar`:

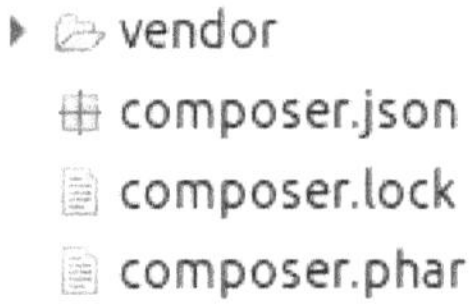

Figure 3.13 – The Composer files and folder

15. The `composer.lock` file is the state file for installed dependencies. The `composer.phar` file is the executable file of Composer, downloaded to be run with the PHP configured in Eclipse.

16. Since the `composer.lock` file is created, you must use the **Update dependencies** button for updating the components installed via Composer:

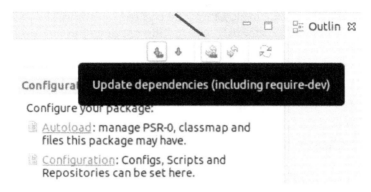

Figure 3.14 – The Update dependencies button

Great! Finally, we have an instance of PHPUnit installed into our project.

> **Tip**
>
> It is recommended that you close and open the project to ensure that autocomplete options will consider the installed dependencies.

So, we have learned how to create a PHP project structure from a use case and how to install the dependencies of a test framework in Eclipse. In the next section, we will start to develop our use case further, aided by tests.

Developing our use case with TDD

From now on, we are going to start TDD. This is an approach popularized by Kent Beck, who is one of the signatories of the *Manifesto for Agile Software Development*. Beck states that TDD is a way of programming without fear. When you write tests before you create the code that they test, you are writing requirements as code. A test that can be run anytime tells us whether code meets the requirements anytime. It is not enough to create a code that works. We need to maintain the code in working order after each change. Tests tell us whether a change has stopped the code from working.

Manifesto for Agile Software Development

In 2001, 17 software developers published a manifesto that exposed their ideas about better ways to develop software. This "Agile Manifesto" presents four main values and twelve principles to guide developers in the management of software construction. The ideas of the Manifesto are the combination of several approaches applied by the authors. These approaches seek to improve the quality of the development software process and the quality of the software as a product, recognizing that change is the only constant. This manifesto is available at `https://agilemanifesto.org`.

In the next section, we will practice one of the approaches of Agile development, creating a test class before creating the class that will be tested.

Creating our SchoolClassTest test class

Let us now create our first test by fulfilling the following steps:

1. In Eclipse, select the `school1` project and access the **File | New | Folder** menu.

2. Create a folder named `test`:

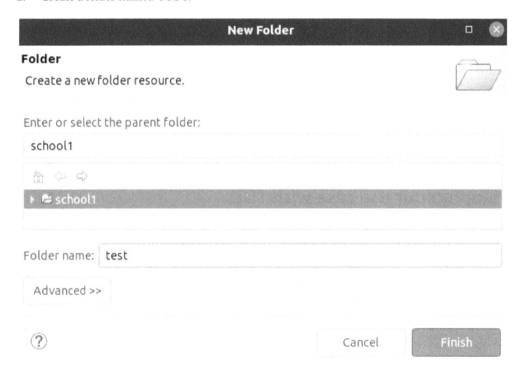

Figure 3.15 – Creating the test folder

3. Select the test folder and create a class using the **File | New | Class** menu. Eclipse will open a window as shown in *Figure 3.16*:

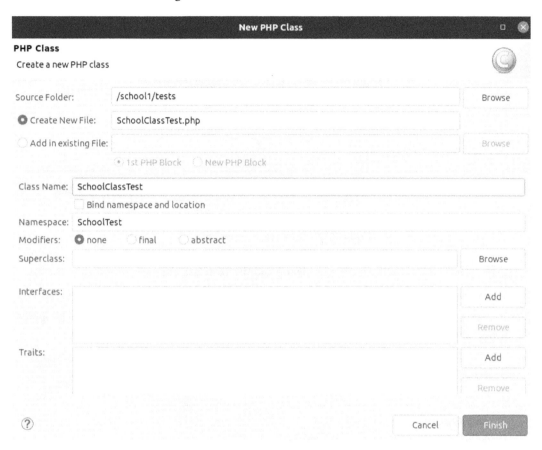

Figure 3.16 – Creating a PHP class

4. Modify the fields in the **New PHP Class** window in the following way:

   - **Class Name**: Set to SchoolClassTest
   - **Bind namespace and location**: Uncheck it
   - **Namespace**: Set to SchoolTest

5. Click on the **Browse** button, which is on the right-hand side of the **Superclass** field. Eclipse will open a window, as shown in *Figure 3.17*:

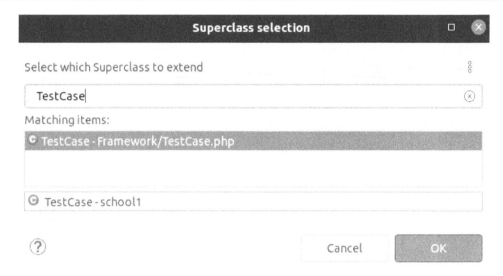

Figure 3.17 – Selecting a Superclass

6. Type TestCase into the text box at the top and the **TestCase** class will be shown in the list.

7. Select **TestCase** and click **OK**.

8. In the next window, **New PHP Class**, click on the **Finish** button. The code editor will open with the following code:

SchoolClassTest.php

```php
<?php
namespace SchoolTest;
use PHPUnit\Framework\TestCase;
class SchoolClassTest extends TestCase
{
}
```

Now, we have the skeleton of our test case class based on PHPUnit. This test case can be run within Eclipse. You can right-click on the test class file and access the **Run As | PHPUnit Test** menu. If you do this now, you will see the following under the **Console** tab:

```
PHPUnit 9.5.10 by Sebastian Bergmann and contributors.
W                                       1 / 1 (100%)
```

This shows that we can run PHPUnit within Eclipse and see the output tests, as we had an open terminal window. Running PHPUnit within Eclipse avoids having to switch between two program windows. Eclipse offers a single panel as an airplane cockpit. W stands for Warning. It means there are no tests.

In this section, we have learned how to create a folder for our tests and how to create a test class extending the PHPUnit TestCase. Specifically, we have created the SchoolClassTest test class. You may have forgotten but we did all that for creating the tests for the SchoolClass business class. In the next section, we will effectively start to practice TDD.

> **What is a business class?**
>
> A class directly related to the customer business can be called a business class. This kind of class has business rules. The software can have many other classes that interact with business classes. Business classes in the MVC pattern are known as models.

Creating a test method in the SchoolClass class

Remember, we call the business class SchoolClass in the list.php file to get a collection of school classes. So, let us create a test to check this. Wait a moment, create a test? But didn't we create the SchoolClass class? Exactly! In TDD, we start with a test that fails, so we don't need to create SchoolClass before writing the test. We only need the class interface, which we have from the class diagram in *Figure 3.5*.

Right, we know that the Schoolclass::getAll method must return an iterable value. So, let's write a test for it using the following steps:

1. Create a method named testListing in the SchoolClass class.

2. Attribute the output of the Schoolclass::getAll method to a variable named $rows.

3. Pass this $rows variable as an argument to the $this->assertIsIterable method.

The SchoolClass test class will look similar to the following code:

SchoolClassTest.php

```php
<?php
namespace SchoolTest;
use PHPUnit\Framework\TestCase;
use School\SchoolClass;
class SchoolClassTest extends TestCase
{
    public function testListing()
    {
        $rows = SchoolClass::getAll();
        $this->assertIsIterable($rows);
```

```
        }
    }
```

The `testListing` method checks whether the result of `SchoolClass::getAll` is iterable through the `assertIsIterable` method. Note that we have added a `use` statement to import the `SchoolClass` class. You can run the test case again. The output under the `Console` tab must be the following:

```
PHPUnit 9.5.10 by Sebastian Bergmann and contributors.
E                                        1 / 1 (100%)
```

E stands for error. Yes, this is not very accurate, but don't worry! There is a tab with more clear data. Click on the **PHPUnit** tab, and you will see a view as shown in *Figure 3.18*:

Figure 3.18 – The PHPUnit tab

The PHPUnit tab shows the trace failure on the right-hand side. In this case, the error message is `Class "School\SchoolClass" not found`. Great! We have a test that fails – the first step of TDD. Now, we will get to the second step – implementing enough code for the test runs.

Creating the SchoolClass class

The error messages are our friends during development because they tell us what we need to do. If the `SchoolClass` class was not found, then we have to find it. For this class to be found, it needs to exist. Let us create this class:

1. Create a folder named `src` in the `schooll` project.

2. Right-click on the `src` folder and create the class by modifying these details:

   - **Class Name**: Set to `SchoolClass`

   - **Bind namespace and location**: Uncheck it

   - **Namespace**: Set to `School`

3. We want to implement the `SchoolClass::getAll` static method. To start, let us write the method interface for this static method:

SchoolClass.php

```php
<?php
declare(strict_types=1);
namespace School;
class SchoolClass
{
    public static function getClass(): iterable
    {
    }
}
```

4. Finally, the moment everyone has been waiting for – we will use the `laminas-db` component to implement the reading of records from the table classes. Do you remember when you installed PHPUnit through Composer? Now, we will use Composer to install `laminas-db`.

5. Double-click on the `composer.json` file and the Composer view will open.

6. Click on the **Require** section and search for `laminas-db`.

7. Select the **laminas/laminas-db** item and click on the **Add Dependencies** button.

8. Press *CTRL + S* and click on the **Update dependencies** button on the top-right-hand side of the Composer view.

Ready! Now, we have a component to access the database.

But how will the `SchoolClass1` class use `laminas-db`? Our business class has to find the classes of the database component of Laminas. You are right. We need to configure the autoloading of our classes so that every class can load another class.

Configuring autoloading of classes

Let us use the following steps for configuring the autoloading of classes:

1. In the **Composer** view, click on the **Autoload** tab. You will see a view similar to the one in *Figure 3.19*.

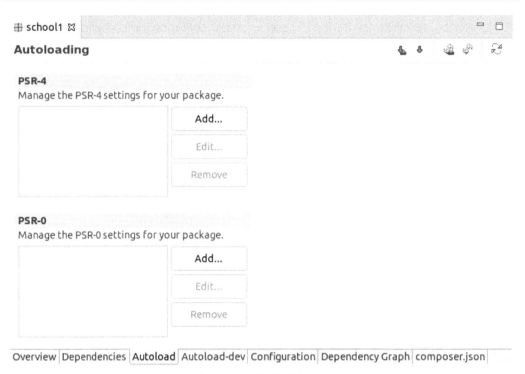

Figure 3.19 – The Autoload tab of the Composer view

The **Autoload** tab has two sections, **PSR-4** and **PSR-0**. These sections are related to PHP-FIG specifications for class autoloading. **PSR-0** is deprecated and exists only for legacy code. We must use **PSR-4** for the new code.

2.   Click on the **Add...** button in the **PSR-4** section. A window named **Edit Namespace** will open (see *Figure 3.20*):

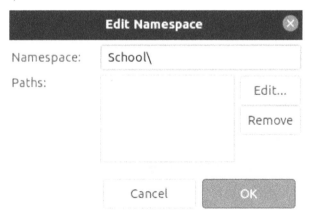

Figure 3.20 – The Edit Namespace window

3. Type School\ in the **Namespace** field and click on the **Edit...** button. A new window, named **Namespace Paths**, will open (see *Figure 3.21*):

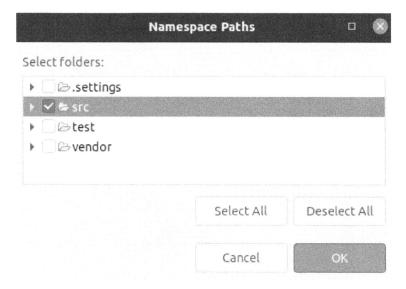

Figure 3.21 – The Namespace Paths window

4. In the **Namespace Paths** window, select the src folder, and click **OK**. Back in the **Edit Namespace** window, click **OK**.

5. What did we do? We have associated the School namespace to the src folder. This means that the classes and interfaces declared with the School namespace will be searched for in the src folder.

You have successfully configured the autoloading of classes related to functional requirements. Let us now do the same for our test classes.

Configuring the autoloading of test classes

Next, we will configure the autoloading for our test classes:

1. In the Composer view, click on the **Autoload-dev** tab.

2. This tab has the same structure as the **Autoload** tab. The difference is that **Autoload-dev** deals with classes used only during development – in other words, classes that won't be packaged for installation in the production environment. Test classes are not necessary in the production environment because they are used only during development and before deployment.

3. In the **PSR-4** section of the **Autoload-dev** tab, click on the **Add...** button.

4. In the **Edit Namespace** window, type SchoolTest\ into the **Namespace** field.

5. Click on the **Edit...** button and select the `tests` folder. Click **OK** and, back in the **Edit Namespace** window, click **OK**.

6. What did we do? We have associated the `SchoolTest` namespace to the `tests` folder. This means classes and interfaces declared with the `SchoolTest` namespace will be searched in the `tests` folder.

7. For these associations to work, it is necessary to regenerate the autoload files.

8. Press *CTRL + S* to save the changes and click on the **Install dependencies** button.

9. No dependency will be installed, but as there was a change in the autoloading configuration, Composer will regenerate the autoload files.

Now, we can use `laminas-db` inside `SchoolClass` without using PHP commands for importing the Laminas classes, such as `include` or `require`.

In the next section, we will implement the configuration file with the database connection data.

Creating the config.php file and rerunning the SchoolClassTest test case

Well, according to our component diagram, `SchoolClass` depends on a file named `config.php`. We will create this file inside a folder named `config`. Create this `config` folder in the root of the `school1` project. Then, inside the `config` folder, create a `config.php` file. The following will be the content of this file:

---

config.php

```php
<?php
return [
    'db' => [
        'driver' => 'Pdo_Mysql',
        'database' => 'school',
        'username' => 'root',
        'password' => '',
        'hostname' => '127.0.0.1'
    ]
];
```

The array inside this file contains the database connection parameters defined in the documentation for `laminas-db`: `https://docs.laminas.dev/laminas-db/adapter/#creating-an-adapter-using-configuration`. The `config.php` file reduces the code changes related to the database. You can change the database server and even the database driver without needing to change the source code of other files. The array defined in `config.php` will be used by the `Adapter` class of Laminas. With this class, we can establish a connection to the database and execute SQL statements. This class implements the **Adapter** design pattern, cataloged by the Gang of Four (Erich Gamma, Richard Helm, Ralph Johnson, and John Vlissides) in the classic book *Design Patterns: Elements of Reusable Object-Oriented Software*. This pattern is useful to allow easy changes between components with different interfaces, such as databases with different drivers. So, we will write the content of the `SchoolClass::getAll` method:

SchoolClass::getAll

```php
public static function getAll(): iterable
{
    $config = require (realpath(__DIR__ . '/../config') .
        'config.php');
    $adapter = new Adapter($config['db']);
    $statement = $adapter->query('SELECT * from classes');
    $resultSet = $statement->execute();
    return $resultSet;
}
```

The `SchoolClass::getAll` method reads the array from the `config.php` file and injects the db key into an instance of the `Adapter` class. With this instance, we prepare a SQL `SELECT` statement and execute it.

> **Note: Automatic code completion**
>
> As you type a class name in the Eclipse code editor, a list of available classes becomes visible. So, if you choose a class from the autocomplete list, Eclipse automatically adds the line of the namespace. But if you don't do that, check whether the namespace is present at the top of the file. For example, the namespace for the Laminas Database Adapter is `Laminas\Db\Adapter\Adapter`.

Now, we can run the `SchoolClassTest` test case again. You will see under the **PHPUnit** tab (see *Figure 3.22*) that the code of the `SchoolClass::getAll` method has passed the test:

Figure 3.22 – The PHPUnit tab showing testListing

As part of TDD, now would be the moment for refactoring and improving the implementation that works. But we assume that our implementation is good enough, and we will create new tests.

Creating a new testInserting test method

We have created an automated test, but it isn't a unit test. As we are connecting to a real database, the `SchoolClassTest->testListing` method is an integrated test: it tests not only the interface of `SchoolClass` but also the connectivity with the database.

Can we test the user interface now, requesting the `index.php` page through the browser? Well, although we have the code to read from the database, there are no records in the `classes` table. So, we don't have school classes in the list. It's better to implement the inserting logic before class creation to test the listing page.

We will add a new test method into `SchoolClassTest`, named `testInserting`, according to the following code snippet:

```php
public function testInserting()
{
    $schoolClass = new SchoolClass(0,'1st year');
    $schoolClass->save();
    $config = require (realpath(__DIR__ . '../config').
        '/config.php');
    $adapter = new Adapter($config['db']);
    $qi = function ($name) use ($adapter) {
        return $adapter->platform->quoteIdentifier($name);
    };
    $fp = function ($name) use ($adapter) {
        return $adapter->driver->formatParameterName($name);
    };
    $statement = $adapter->query('SELECT name from classes
```

```
            where ' . $qi('name') . ' = ' . $fp('name'));
    $resultSet = $statement->execute(['name' => '1st
            year']);
    $this->assertCount(1,$resultSet);
    $statement = $adapter->query('DELETE from classes where
            ' . $qi('name') . ' = ' . $fp('name'));
    $resultSet = $statement->execute(['name' => '1st
            year']);
    }
```

The `SchoolClassTest->testInsertingList` method checks whether `SchoolClass->save` produces a new record for the `classes` table. The test method deletes the record for restarting the data state. This test method presumes, of course, that the `name` field of the table classes prevents duplicate values. So, there must be only one record called `'1st year'`. As part of the first step of TDD, if you run this test now, it will fail. The reason is obvious: the `SchoolClass->save` method doesn't exist. Let us implement it!

> **Note: Closures for treating field names and values**
>
> In the `SchoolClassTest->testInsertingList` method, we use the `$qi` and `$fp` closures for quoting the field name and formatting the value respectively. So, if we change our database from MySQL to PostgreSQL, for instance, it won't be necessary to change the code into a class but only to change the connection parameter in the `config.php` file. In addition, formatting the values appropriately avoids SQL injection attacks.

According to our class diagram, the `SchoolClass->save` method returns a Boolean value. If the method inserts successfully, it returns true – otherwise, it returns false. The code of this method is as follows:

```
public function save(): bool
{
    $adapter = self::getAdapter();
    $fp = function ($name) use ($adapter) {
      return $adapter->driver->formatParameterName($name);
    };
        $statement = $adapter->query('INSERT INTO
            classes(name) VALUES (' . $fp('name') . ')');
    try {
      $statement->execute(['name' => $this->name]);
    } catch (\Exception $e) {
```

```
        error_log($e->getMessage());
        return false;
    }
    return true;
}
```

To insert a new record, we need to connect to a database. So, we have extracted the database adapter creation code to a separate `SchoolClassTest::getAdapter` static method:

```
private static function getAdapter(): AdapterInterface
    {
        $config = require (realpath(__DIR__ . '/../config') .
            '/config.php');
        $adapter = new Adapter($config['db']);
        return $adapter;
    }
```

This way, we can avoid repeating this code for every method that needs to connect to the database. Of course, we had to refactor the `SchoolClassTest::getAll` method to use `SchoolClassTest::getAdapter`:

```
public static function getAll(): iterable
    {
        $adapter = self::getAdapter();
        $statement = $adapter->query('SELECT * from classes');
        $resultSet = $statement->execute();
        return $resultSet;
    }
```

Coming back to the `SchoolClassTest->save` method, you should notice that it executes SQL INSERT using an `$this->name` attribute. So, this attribute must be defined and have a value. In fact, we need to add two attributes to the class, one for each related field from the `classes` table:

```
public int $code = 0;
public string $name = '';
```

> **Why use public attributes in a class?**
>
> There is a recipe for creating classes that state you must create private attributes and public methods. So, if you need to read or write an attribute, you will need to create a method for that. If you have twenty attributes and need to read and write all of them, you will need to create forty methods. The question is: is it really necessary? If you don't transform the data attribute before reading or after writing, you are losing time using a method only to do what you can do directly.

The `$code` and `$name` attributes must be populated when the instance of the `SchoolClass` class is created. So, we need to declare the constructor method:

```php
public function __construct(int $code = 0, string $name = '')
{
    $this->code = $code;
    $this->name = $name;
}
```

Now, you can run the test case successfully. We get to list and insert the school classes.

Creating an edition form edit.php

It's time to create the edition form in the `edit.php` file. Why? Because we need a user interface to add school classes. Input the following code into the project root by navigating through **File** | **New** | **PHP File**. The following source code is the content of `edit.php`:

edit.php

```php
<!DOCTYPE html>
<html>
<head>
<meta charset="UTF-8">
<title>Classes Registration</title>
</head>
<body>
    <h1>Classes Registration</h1>
    <form action="save.php" method="post">
    Name: <input type="text" name="name" autofocus=
        "autofocus">
    <br/>
        <input type="hidden" name="code"><br/>
```

```
        <input type="submit" value="save">
    </form>
<p>
    <a href="index.php">Homepage</a>
</p>
</body>
</html>
```

This web page will produce a page as shown in *Figure 3.23*:

Classes Registration

Name:

save

Homepage

Figure 3.23 – The form rendered from the edit.php file

As you should perceive, this web form won't request successfully because the save.php file doesn't exist. Let's create it then:

---

save.php

```php
<?php
use School\SchoolClass;
require 'vendor/autoload.php';
$name = ($_POST['name'] ?? null);
$schoolClass = new SchoolClass(0,$name);
$schoolClass->save();
header('Location: index.php');
```

The content of save.php is very simple (after we have created the test for inserting). The file gets the value of the field name from the HTML form, sent via HTTP POST, and passes it to an instance of SchoolClass to persist the data into the classes table. In the next step, the file makes an HTTP request with a code of 302 through the header PHP function, which redirects to the index.php page.

Well, you can now insert a school class using the form in edit.php. The list will be void. Don't worry! It happens because we have to import the autoload.php file of Composer. It is enough to add the following line after the use statement in list.php:

```
require 'vendor/autoload.php';
```

After adding that line, you can refresh the index.php page, and you will see a page as follows:

Classes Registration

Add a new class

code name
<u>1</u>   9th year <u>Remove</u>

Figure 3.24 – A list of school classes

Great! We can insert and list the school classes through the web interface. Now, we will implement the updating action. Of course, we will start with a test.

Creating a SchoolClassTest->testUpdating test method

The following source code shows the SchoolClassTest->testUpdating test method:

```
public function testUpdating()
{
    $schoolClass = new SchoolClass(0,'1st year');
    $schoolClass->save();
    $schoolClass = SchoolClass::getByName('1st year');
    $schoolClass->name = '2nd year';
    $schoolClass->save();
    $schoolClass = SchoolClass::getByName('2nd year');
    $this->assertEquals('2nd year', $schoolClass->name);
    $schoolClass->delete();
    $schoolClass = SchoolClass::getByName('2nd year');
    $this->assertEmpty($schoolClass->name);
}
```

This test tries to insert a new school class and then change the name of this school class. As part of the first step of TDD, if you run this test now, it will fail. Right, this is expected. If the test passes at this point, something is very wrong indeed!

You should perceive that we are using a `SchoolClass->delete` method to remove the school class from the table this time. This method was planned in the class diagram and it is better to use it than to repeat several lines.

Next, let us run the test successfully. You should have noticed that the `SchoolClassTest->testUpdating` method calls a `SchoolClass::getByName` method. This method doesn't exist, so we have to create it. Let us create this method after the class constructor:

```
public static function getByName($name = null): SchoolClass
{
  $adapter = self::getAdapter();
  $qi = function ($name) use ($adapter) {
    return $adapter->platform->quoteIdentifier($name);
  };
  $fp = function ($name) use ($adapter) {
    return $adapter->driver->formatParameterName($name);
  };
  $statement = $adapter->query('SELECT * from classes
      WHERE ' . $qi('name') . '=' . $fp('name'));
  $resultSet = $statement->execute(['name' => $name]);
  if ($resultSet->count() == 0){
    return new SchoolClass();
  }
  $row = $resultSet->current();
  return new SchoolClass((int)$row['code'],$row['name']);
}
```

The `SchoolClass::getByName` method is similar to the `SchoolClass::getAll` method, at least until the statement creation. SQL `SELECT` filters by name and if no record is found, a void instance of `SchoolClass` will be returned. Is this enough to run the test successfully? Well, you can run the test and discover the answer – no!

When we implemented the `SchoolClass->save` method, the method was only prepared to insert records, not to update them. So, even if you change the name of an object of the `SchoolClass` class, when you call the `SchoolClass->save` method, it will try to insert the record because it is the only code there. So, we have to change this method so it can perform updates too. The method will be as follows:

```
public function save(): bool
{
```

```php
$adapter = self::getAdapter();
$fp = function ($name) use ($adapter) {
  return $adapter->driver->formatParameterName($name);
};
if (empty($this->code)){
  $statement = $adapter->query('INSERT INTO
      classes(name) VALUES (' . $fp('name') . ')');
  $params = ['name' => $this->name];
} else {
  $statement = $adapter->query('UPDATE classes SET
      name=' . $fp('name') . ' WHERE code=' .
          $fp('code'));
  $params = [
    'code' => $this->code,
    'name' => $this->name
  ];
}
try {
  $statement->execute($params);
} catch (\Exception $e) {
  error_log($e->getMessage());
  return false;
}
return true;
}
```

The code for the school class is autogenerated by the table, so we don't expect an insertion. If the code has a value, we assume that is an update action. Will the test work now? No, because the `SchoolClass->delete` method doesn't exist yet. Don't waste time – create it:

```php
public function delete(): bool
{
  $adapter = self::getAdapter();
  $fp = function ($name) use ($adapter) {
    return $adapter->driver->formatParameterName($name);
  };
  $statement = $adapter->query('DELETE FROM classes WHERE
```

```
        code=' . $fp('code'));
    try {
      $statement->execute([
        'code' => $this->code
      ]);
    } catch (\Exception $e) {
      error_log($e->getMessage());
      return false;
    }
    return true;
}
```

Now, you can run the test successfully. Finally, we get out of a red state and into a green state. And the current implementation of SchoolClass allows us to alter the edit.php and save.php files to change the school classes through the web interface.

Altering edit.php

Let's start with edit.php:

---

edit.php

```php
<?php
use School\SchoolClass;
require 'vendor/autoload.php';
$code = ($_GET['code'] ?? null);
$schoolClass = SchoolClass::getByCode($code);
?>
<!DOCTYPE html>
<html>
    <head>
    <meta charset="UTF-8">
        <title>Classes Registration</title>
    </head>
<body>
    <h1>Classes Registration</h1>
    <form action="save.php" method="post">
    Name: <input type="text" name="name"
```

```
            autofocus="autofocus"
        value="<?=$schoolClass->name?>"><br/>
            <input type="hidden" name="code" value="<?=
                $schoolClass->code?>"><br/>
        <input type="submit" value="save">
        </form>
    <p>
    <a href="index.php">Homepage</a>
    </p>
    </body>
    </html>
```

What is the difference between the previous version and the current version of this file? The answer is the PHP block code, with the import of the `autoload.php` files, the reading of the parameter code (sent via `HTTP GET`), and the recovery of an instance of `SchoolClass` from the code. Wait a moment! Where did this `SchoolClass->getByCode` method come from? You are very smart! It doesn't exist yet! We have to implement it! Let's go:

```php
public static function getByCode($code = null): SchoolClass
{
    $adapter = self::getAdapter();
    $qi = function ($name) use ($adapter) {
        return $adapter->platform->quoteIdentifier($name);
    };
    $fp = function ($name) use ($adapter) {
        return $adapter->driver->formatParameterName($name);
    };
    $statement = $adapter->query('SELECT * from classes
        WHERE ' . $qi('code') . '=' . $fp('code'));
    $resultSet = $statement->execute(['code' => $code]);
    if ($resultSet->count() == 0){
        return new SchoolClass();
    }
    $row = $resultSet->current();
    return new SchoolClass((int)$row['code'],$row['name']);
}
```

Yes, the `SchoolClass->getByCode` method is very similar to the `SchoolClass->getByName` method. It's good if you've noticed that. Don't worry because we will eliminate the repetition later. But now we need to focus on updating the school classes through the web interface. Since the page with the edit form was changed, let's change the `save.php` file, which is in charge of the persistence.

Altering save.php

This file will be coded as follows:

```php
<?php
use School\SchoolClass;
require 'vendor/autoload.php';
$code = ($_POST['code'] ?? null);
$name = ($_POST['name'] ?? null);
$schoolClass = new SchoolClass((int)$code, $name);
$schoolClass->save();
header('Location: index.php');
```

Observe that we have only changed the incoming parameter code. The value of this parameter is what determines whether the operation is an insert or an update. With this change, you can now alter the name of inserted school classes. Try to insert several school classes and then, change the names of some of them.

And now, the moment everyone was waiting for – the removal of school classes.

Creating the delete.php file

According to our component diagram and the hyperlink in the `list.php` file, the removal operation has to be inside the `delete.php` file. Let us create it:

delete.php

```php
<?php
use School\SchoolClass;
require 'vendor/autoload.php';
$code = ($_GET['code'] ?? null);
$schoolClass = SchoolClass::getByCode((int)$code);
$schoolClass->delete();
header('Location: index.php');
```

Remember, we already implemented the `SchoolClass->delete` method and it was tested. Now, you can insert, edit, delete, and list school classes through the web interface. You can start the PHP-embedded web server in the `school1` folder as follows:

```
php -S localhost:8000
```

Access the `localhost:8000` address and play with the school classes. The use case of managing school classes is implemented. But, as we are working with TDD, it is important to ask whether all our code is covered by tests. We can find out by generating a report called code coverage through PHPUnit.

Generating code coverage reports

Let's automate the command that produces the code coverage report with the **External Tools Configurations** of Eclipse. You made a configuration of this type before, in *Chapter 2, Setting Up the Environment for Our E-Commerce Application*, remember? Let us do it again:

1. Access the **Run | External Tools/External Tools Configurations....** menu as you did before and create a configuration with the following parameters under the **Main** tab:

    *   **Name**: `Code Coverage`

    *   **Location**: `${project_loc}/vendor/bin/phpunit`

    *   **Working Directory**: `${project_loc}/src`

    *   **Arguments**: `--coverage-html code_coverage`

2. Don't forget to check the **External Tools** item under the **Common** tab:

Figure 3.25 – The checkbox for displaying the external tools in the menu

3. After saving the configuration, click **Apply** and **Close**.

4. To generate the code coverage report, we need to create a configuration file for PHPUnit. I promise that this is the last time we need to use the terminal.

5. Inside the directory, `school1` in the workspace, type the following command:

    ```
    vendor/bin/phpunit --generate-configuration
    ```

6. You will choose the default option for each of the questions presented by PHPUnit. The output will be as follows:

    ```
    Bootstrap script (relative to path shown above; default:
    vendor/autoload.php):
    Tests directory (relative to path shown above; default:
    tests):
    Source directory (relative to path shown above; default:
    src):
    Cache directory (relative to path shown above; default:
    .phpunit.cache):
    Generated phpunit.xml in /opt/lampp/htdocs/school1.
    Make sure to exclude the .phpunit.cache directory from
    version control.
    ```

7. This command will generate a file named `phpunit.xml`. Edit this file and change this parameter:

    ```
    forceCoversAnnotation="true"
    ```

    Change it to the following:

    ```
    forceCoversAnnotation="false"
    ```

8. Create a folder named `code_coverage` inside the `school1` project.

9. Finally, select the `school1` project and run the new **Code Coverage** configuration on the toolbar. You can follow the generation of the report under the **Console** tab of Eclipse, which will end with this message:

    ```
    Generating code coverage report in HTML format ... done
    [00:00.013]
    ```

10. You will find a `index.html` file. You can open it with the internal browser of Eclipse, selecting the file and accessing the **Open With | Web Browser** pop-up menu. You should see the following screen:

/opt/lampp/htdocs/school1/src / (Dashboard)

Code Coverage

	Lines		Functions and Methods		Classes and Traits	
Total	69.35%	43 / 62	57.14%	4 / 7	0.00%	0 / 1
⟨⟩	69.35%	43 / 62	57.14%	4 / 7	0.00%	0 / 1
SchoolClass.php						

Legend

Low: 0% to 50%   **Medium**: 50% to 90%   **High**: 90% to 100%

Generated by php-code-coverage 9.2.10 using PHP 7.4.3 and PHPUnit 9.5.10 at Tue Dec 21 2:10:01 UTC 2021.

Figure 3.26 – The dashboard for code coverage

11. Observe that our `SchoolClass` class is not fully covered with tests. What is missing? The report can answer this question.

12. Click on the **SchoolClass.php** link. It will open a page with details about the code coverage, as we can see in *Figure 3.27*:

	Classes and Traits		Functions and Methods			Lines	
Total	0.00%	0 / 1	57.14%	4 / 7	CRAP	69.35%	43 / 62
SchoolClass	0.00%	0 / 1	57.14%	4 / 7	16.14	69.35%	43 / 62
construct			100.00%	1 / 1	1	100.00%	3 / 3
getByCode			0.00%	0 / 1	6	0.00%	0 / 13
getByName			100.00%	1 / 1	2	100.00%	13 / 13
getAll			100.00%	1 / 1	1	100.00%	4 / 4
save			0.00%	0 / 1	3.07	80.00%	12 / 15
delete			0.00%	0 / 1	2.08	72.73%	8 / 11
getAdapter			100.00%	1 / 1	1	100.00%	3 / 3

Figure 3.27 – Detailed code coverage of SchoolClass.php

The page about `SchoolClass.php` shows that the `SchoolClass->getByCode` method has no coverage and the `SchoolClass->save` and `SchoolClass->delete` methods are not totally covered. If you scroll down the page, you can see all the source code colored according to the code coverage. We know that the `SchoolClass->getByCode` method is not covered by a test because we didn't create one. The problem with the `SchoolClass->save` and `SchoolClass->delete` methods is that parts related to exceptions are never executed by tests.

The `SchoolClass->getByCode` method can be covered by a test simply by changing the `SchoolClassTest->testInserting` test method. So, let us write the following code:

```
public function testInserting()
{
    $schoolClass = new SchoolClass(0,'1st year');
    $schoolClass->save();
    $schoolClass = SchoolClass::getByName('1st year');
    $this->assertEquals('1st year',$schoolClass->name);
    $code = $schoolClass->code;
    $schoolClass = SchoolClass::getByCode($code);
    $this->assertEquals('1st year',$schoolClass->name);
    $schoolClass->delete();
    $schoolClass = SchoolClass::getByCode($code);
    $this->assertEmpty($schoolClass->name);
}
```

We won't create tests for exceptions because doing so is not currently relevant. We already have good test coverage for our code, but the most important part is that we have learned how to do it with the help of Eclipse, with just a click. We have also learned about how to use Composer inside Eclipse, increasing our knowledge about integration between Eclipse and a tool for PHP development.

Summary

In this chapter, we have learned about how to use an isolated Laminas component. We designed a registration for school classes, installed PHPUnit, integrated it with Eclipse, installed `laminas-db`, and connected it to the `school` database. So, we can see that is possible to use Laminas components without the MVC core. Specifically, the example project, `school1`, facilitated our learning about how to abstract database connections with `laminas-db`. The `school1` project showed us how a PHP application can use Laminas as a library, calling the framework components. This is a scenario where your application has control of the main execution flow.

At the same time, we followed the good practice of creating tests before creating the code to be tested. This is very important because we will always start with tests in the following chapters. So, we will combine the power of Laminas components with the advantages of TDD.

In the next chapter, we will refactor the `school1` project as one new project to advance one step further with the use of the `laminas-db` component. The `school1` project is fully available at `https://github.com/PacktPublishing/PHP-Web-Development-with-Laminas/tree/main/chapter03/school1`.

4

From Object-Relational Mapping to MVC Containers

This chapter continues to prepare you to use Laminas. Here, we will go through the last two stages:

- The handling of databases with **object-relational mapping** (**ORM**), abstracting SQL statements
- The inversion of control for object-oriented PHP applications with the Laminas framework as a container

First, we will explain the foundations of ORM implemented with the aid of `laminas-db`. Then, we will introduce you to the basics of Laminas by building an MVC application. These foundations are very important for the creation of an e-commerce application and learning them will be invaluable. You could jump to *Chapter 5, Creating the Virtual Store Project*, but you may feel like you are trying to attack the Death Star in a hang glider. Keep calm and prepare yourself for battle.

In this chapter, we'll be covering the following topics:

- Using ORM
- Use case – managing students
- Using Laminas as a container

Technical requirements

All the code related to this chapter can be found at `https://github.com/PacktPublishing/PHP-Web-Development-with-Laminas/tree/main/chapter04`.

Using ORM

You have learned how to connect to a relational database with `laminas-db` using its implementation of the **Adapter** design pattern. However, although we created a structure that allows switching between different SQL database vendors through configuration, we kept writing SQL statements. When we write SQL statements directly inside the code of an application, we have to choose a specific SQL dialect. Although there is a SQL standard, the fact is that there are some differences, such as the use of quotations and the names of functions.

For example, in MySQL, a query on the `resources` table of the `whatstore` database would have this form:

```
SELECT `code`,`name`,`method` FROM `resources`
```

On the other hand, the same query in PostgreSQL would look like this:

```
SELECT "code","name","method" FROM "public"."resources"
```

As we can see, the quotation character is different for each one of these vendors. If you use the `SELECT` query of one of them in the other, an error will occur. Imagine changing several files of an application because you are migrating from one vendor to another.

You might say that some string replacement based on a regular expression could solve this fast. However, in addition to the quotation issue, there are differences between database functions. For example, MySQL aggregates values of fields from several table records, like this:

```
SELECT GROUP_CONCAT(`name`) as `names` FROM `resources`
```

But PostgreSQL uses another function for the same result:

```
SELECT string_agg("name",',') AS "names" FROM
"public"."resources"
```

So, database migration can be a hard task if the application directly writes the SQL statements.

ORM avoids the rewriting of code if you need to migrate from one database vendor to another. This facility is possible thanks to SQL statement encapsulation.

To get started with ORM, we will learn how to deal with tables as objects. The `laminas-db` component implements the **Table Data Gateway** and **Row Data Gateway** design patterns, which allow access to a table or a record set as an object and allow access to a single record as an object, respectively. In this chapter, we will use the Table Data Gateway pattern.

Table Data Gateway

Martin Fowler, in his book *Patterns of Enterprise Application Architecture*, *Addison-Wesley*, defines Table Data Gateway as "an object that acts as a Gateway to a database table." This object is intended to handle several rows from a table. Gateway is also the name of a design pattern. Martin Fowler, in the same book, defines a Gateway as "an object that encapsulates access to an external system or resource." Databases are external systems for web applications, so it makes sense that some object-relational patterns extend the gateway pattern. You can search for one single record with a Table Data Gateway, but this object always returns a record set (in this case, a set with one only element).

Row Data Gateway

Martin Fowler, in his book *Patterns of Enterprise Application Architecture*, *Addison-Wesley*, defines Row Data Gateway as "an object that acts as a Gateway to a single record in a data source." While Table Data Gateway focuses on a record set, Row Data Gateway focuses on a single record.

We will create a `school2` project in this chapter, which will implement ORM with `laminas-db`. This `school2` project has a difference, related to the separation of responsibilities, from the `school1` project. In `school1`, the same class (`SchoolClass`) stores the business data and deals with communication with the database. Now, we will have a class to store the data in memory that doesn't know the data source and another class to read and write to the database. This approach avoids coupling the business rules with the data storage model. We can separate responsibilities in the following way:

1. Make a copy of the `school1` project in the `htdocs` folder and name it `school2` in the same `htdocs` folder. Remember that we agreed in *Chapter 2*, *Setting Up the Environment for Our E-Commerce Application*, that the `htdocs` folder would be our workspace throughout the book. You can do this by selecting the `school1` project with the right mouse button and clicking on **Copy**, or simply pressing *Ctrl + C*. Then, press *Ctrl + V* and you will see a dialog box, as in *Figure 4.1*:

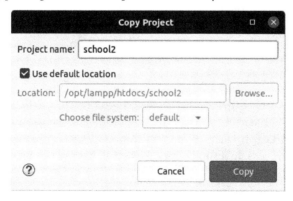

Figure 4.1 – Dialog box for copying a project

2. Make sure the **Project name** field is filled with `school2`; then, click on **Copy**.

3. Select the `school1` project with the right mouse button and click on the **Close Project** item.

We are closing our first project because we don't need it anymore and if it stays open, we may accidentally change a file in that project thinking it belongs to the `school2` project. From now on, we will work on the `school2` project.

Use case – managing students

In this second project (`school2`), we will implement one more use case, *managing students*. This use case is very similar to *managing school classes*. Both have four flows, related to create, recover, update, and delete actions. The difference is that the `student` model is dependent on the `school class` model. So, a precondition for the flows of the *managing students* use case is that the `classes` table must have at least one record.

Implementing the use case

By now, you know how to create folders. So, let us implement a use case by following these steps:

1. Create a new folder named `view` in the root of the project.

2. Inside this folder, create two more folders: `schoolclass` and `student`.

    The `view` folder is an explicit division for files related to the user interface. This organization will help us to identify what belongs to the presentation layer.

3. Next, select the following files:

    - `delete.php`

    - `edit.php`

    - `index.php`

    - `list.php`

    - `save.php`

4. Copy these files to the `view/schoolclass` and `view/student` folders.

5. Now, you can remove all of them from the project root, except for the `index.php` file. The `index.php` file is the starting point of the application and will act as a coordinator for the other files. The `index.php` file is akin to the front door of a house. You need to get into the house to access the rooms. In a similar way, a user will have to use `index.php` before they access the other PHP files.

6. Replace the content of this `index.php` file with the following source code:

index.php

```
<!DOCTYPE html>
<html>
<head>
<meta charset="UTF-8">
<title>School Management</title>
</head>
<body>
  <h1>School Management</h1>
  <ul>
  <li><a href="view/schoolclass">School classes
        </a></li>
  <li><a href="view/student">Students</a></li>
  </ul>
</body>
</html>
```

As you can see, the home page is a static page that has links for the two registrations. At this stage, the `index.php` file is purely presentational. Now, we will refactor the other presentation files. Let's start with the use case we have implemented in the `school1` project, *managing school classes*.

We need to change the path for importing the `autoload.php` file in the files inside the `view/schoolclass` and `view/student` folders. As the files inside these folders were in the application root folder before, the path is now wrong. Inside the `view/schoolclass` and `view/student` folders, all files that import `autoload.php` must have, from now on, the following line at the beginning:

```
require realpath(__DIR__ . '/../../vendor') .'/autoload.php';
```

If you don't make this change, you won't be surprised by an error related to this path. The complete source code is provided, but it's good practice to code instead of copying and pasting. When you type, you can make mistakes and need to review the code you have typed to avoid errors. This process helps to assimilate the content. It is very important to know what you are doing; otherwise, you will only be a "code operator."

These first steps give us the general outline of our new project structure. We have two registrations, one for school classes and another for students. Let's start by refactoring the code for the school classes, which we have inherited from the `school1` project. Remember that test classes remain in the same place.

Now that we have implemented the use case, let us run our first test.

Running the SchoolClassTest->testListing test

We will create a test to check whether a list of school classes is generated successfully. This test will be implemented by the `SchoolClassTest->testListing` method:

1. The `SchoolClassTest->testListing` method looks as follows:

    ```
    public function testListing()
    {
        $rows = SchoolClassTable::getInstance()->getAll();
        $this->assertIsIterable($rows);
    }
    ```

    Of course, this test will fail because the `SchoolClassTable` class doesn't exist yet. This class is an implementation of the *Table Data Gateway* pattern using `laminas-db`. The `SchoolClassTable` instance is not created directly with a new operator, but it is provided by an implementation of the **Singleton** pattern, which limits the number of instances for a class. This pattern is very useful when you know that only one instance of a class is necessary for a specific context, and you don't want a developer to create another instance accidentally after implementing a new feature or fixing a bug. The *Singleton* pattern needs a private constructor, a static attribute for the single instance, and a static method for creating and recovering this instance.

    The `SchoolClassTable` class must stay in the `src` folder.

    > **Note**
    >
    > From now on, we won't talk about things we already know, such as creating classes with Eclipse.

2. Create the `SchoolClassTable` class in the `src` folder, always with the `School` namespace. This class needs to have an instance for the data table gateway of school classes. Hence, we will implement an `$instance` attribute for `SchoolClassTable` like this:

    ```
    private static ?SchoolClassTable $instance = null;
    ```

    The question mark before the type of the `$instance` attribute allows us to initialize it with `null`. We postpone object creation until it is required.

3. The constructor of `SchoolClassTable` looks as follows:

    ```
    private function __construct(TableGatewayInterface
        $tableGateway)
    {
    ```

```
        $this->tableGateway = $tableGateway;
    }
```

Observe that we are injecting an instance of `TableGatewayInterface` into a `$this->tableGateway` attribute. `TableGatewayInterface` is an interface for *Table Data Gateway* implementations. It is good practice to use interfaces instead of classes in dependency injection contexts because it makes it easy to replace the implementations. We will use here the Laminas implementation for *Table Data Gateway*, but the interface allows you to change the implementation to another without changing the `SchoolClassTable` class. The changes are inevitable; however, we can reduce the number of changes and the need for rewriting code.

4. Finally, to complete the **Singleton** pattern, we will have a `SchoolClassTable::getInstance` method:

```
    public static function getInstance():
        SchoolClassTable
    {
      if (null == self::$instance) {
        $adapter = self::getAdapter();
        $resultSetPrototype = new ResultSet();
        $resultSetPrototype->setArrayObjectPrototype
            (new SchoolClass());
        $tableGateway = new TableGateway('classes',
            $adapter, null, $resultSetPrototype);
        self::$instance = new
            SchoolClassTable($tableGateway);
      }
      return self::$instance;
    }
```

The `TableGateway` class of `laminas-db` requires the injection of an `AdapterInterface` instance. This instance is provided by the `SchoolClassTable::getAdapter` method that we have brought from the former `SchoolClass` class of the `school1` project.

Know that although we are using a database adapter implementation of `laminas-db`, you can replace it with anything that implements `AdapterInterface`. This is an essential feature of Laminas: you can replace most components with others that are compatible or write your own component. You have the convenient option of delegating a task to a ready component, but you also have the power to take control absolutely, by implementing customized code. Of course, you can always write your own code in open source software, but in reality, it is sometimes difficult to connect your components to existing ones.

You may be curious about the fourth parameter of the `TableGateway` constructor:

```
$tableGateway = new TableGateway('classes', $adapter,
null, $resultSetPrototype);
```

The first parameter is the table name. The second parameter is the database adapter object. The third parameter, which is optional, allows us to inject an object that extends the table gateway features. The fourth parameter is also optional, but if provided allows us to encapsulate the elements of a result set into instances of a model class. Usually, the return of a SQL SELECT statement executed by `TableGateway` will be a collection of `ArrayObject`, a PHP native class. But the `ResultSet` class, through the `setArrayObjectPrototype` method, allows us to define the type of result collection. In this case, we defined that queries made by `SchoolClassTable` will return a collection of `SchoolClass` instances.

Right, by this point, we have solved the absence of `SchoolClassTable`. But the `SchoolClassTest->testListing` test keeps failing. This failure occurs because the test expects that the `SchoolClassTable` instance has a `getAll` method. At this moment, we will avoid duplicate work. As both the use cases have similarities, do you agree that they share common code? The `getAll` method is an example. The `StudentTable` class, which we will have to implement for the *managing students* use case, also needs a `getAll` method.

When one or more classes have duplicated code, it's time to create a parent class that encapsulates the common code.

5. So, we will create an `AbstractTable` class in the `src` folder. This class must implement the `getAll` method like this:

```
public function getAll(): iterable
{
    return $this->tableGateway->select(null);
}
```

This method is the object-oriented dream: only one line of code. The method comprises total outsourcing. What does this mean? The `getAll` method does not do anything directly. The `getAll` method has to call another method. Instead of writing SQL SELECT by ourselves, we let the `TableGateway` instance write it through its `select` method. We don't need to write the SQL SELECT statement or worry about quotation marks.

`AbstractTable` will keep the common code for Table Data Gateway implementations of our project; `AbstractTable` has also declared the attribute for the `TableGateway` instance:

```
protected TableGatewayInterface $tableGateway;
```

It's a `protected` attribute because it must be visible only to the children of `AbstractTable`, which will deal with it.

We have established that `SchoolClassTable` must extend `AbstractTable` and that database adapter creation is a common feature for both Table Data Gateway implementations, so the static `getAdapter` method must be in the `AbstractTable` class.

Now, the `SchoolClassTest->testListing` method will run successfully. But the other test methods will fail. Of course, we need to refactor them too.

6. Let's refactor the `SchoolClassTest->testInserting` method. This test method covers inserting, selecting, and deleting actions. We insert a record, recover it, and delete it:

```php
public function testInserting()
{
    $schoolClassTable = SchoolClassTable::
        getInstance();

    $schoolClass = new SchoolClass(0,'1st year');
    $schoolClassTable->save($schoolClass);

    $schoolClass = SchoolClassTable::getInstance()
        ->getByField('name', '1st year');
    $this->assertEquals('1st year',
        $schoolClass->name);
    $code = $schoolClass->code;
    $schoolClass = SchoolClassTable::
        getInstance()->getByField('code', $code);
    $this->assertEquals('1st year',
        $schoolClass->name);
    $schoolClassTable->delete($schoolClass->code);
    $schoolClass = SchoolClassTable::
        getInstance()->getByField('code', $code);
    $this->assertEmpty($schoolClass->name);
}
```

If you try to run the `SchoolClassTest` test case now, it will fail because the `SchoolClassTable` class doesn't have the `getByField` method. This method will replace the `getByCode` and `getByName` methods of the former `SchoolClass` class of the `school1` project.

7. The `getByField` method, however, can be implemented by the `AbstractTable` class because it is generic:

```php
public function getByField(string $field, $value):
    AbstractModel
{
    $where = [
        $field => $value
    ];
    $rowSet = $this->tableGateway->select($where);
    if ($rowSet->count() == 0) {
        $modelName = $this->modelName;
        return new $modelName();
    }
    return $rowSet->current();
}
```

The `getByField` class allows searching for any field. As you have noticed, if no record is found, the method returns an instance of a class defined by the `$this->modelName` attribute. In the `SchoolClassTable` class, this attribute is declared as follows:

```php
protected string $modelName = 'School\SchoolClass';
```

This attribute allows each child class of `AbstractTable` to inform its `model` class and ensures that the method will return an instance of `AbstractModel`.

Wait! What is `AbstractModel`? It's a generalization for our models. It implements two generic methods, one for data input, `exchangeArray`, and the other for data output, `toArray`.

This is the model:

AbstractModel.php

```php
<?php
declare(strict_types = 1);
namespace School;

abstract class AbstractModel
{

    public function exchangeArray($data)
    {
```

```
    $attributes = get_object_vars($this);
    foreach ($attributes as $attribute => $value) {
      $this->$attribute = (is_int($this->$attribute) ?
        (int) $data[$attribute] : $data[$attribute]);
    }
  }


  public function toArray()
  {
    return get_object_vars($this);
  }
}
```

The exchangeArray method receives a data array and puts the values in the respective attributes. The toArray method, on the other hand, returns an array with the values of the attributes.

Will the test run successfully now? Of course not!

8. We have to implement the save method in the AbstractTable class:

```
public function save(AbstractModel $model): bool
{
  $set = $model->toArray();
  $keyName = $this->keyName;
  try {
    if (empty($model->$keyName)) {
      unset($set[$keyName]);
      $this->tableGateway->insert($set);
    } else {
      $this->tableGateway->update($set, [
        $keyName => $set[$keyName]
      ]);
    }
  } catch (\Exception $e) {
    error_log($e->getMessage());
    return false;
  }
  return true;
}
```

The `save` method delegates the hard work to the `insert` and `update` methods of `TableGateway`. The logic is the same as in the former `SchoolClass` class of the `school1` project: the choice between an insert and an update depends on the value of the primary key generated by the table. Observe that the name of the primary key field is given by the `$this->keyName` attribute. This attribute must be declared by the `AbstractTable` class, as follows:

```
protected string $keyName;
```

9. The `SchoolClassTest` test case won't run yet because we have to implement the `delete` method in the `AbstractTable` class:

```php
public function delete($value): bool
{
  try {
    $this->tableGateway->delete([
      $this->keyName => $value
    ]);
  } catch (\Exception $e) {
    error_log($e->getMessage());
    return false;
  }
  return true;
}
```

Now will the `SchoolClassTest` test case run successfully? Not yet.

10. You have to create the `SchoolClass` class first, as follows:

SchoolClass.php

```php
<?php
declare(strict_types = 1);
namespace School;

class SchoolClass extends AbstractModel
{

  public int $code = 0;

  public string $name = '';
```

```php
    public function __construct(int $code = 0, string
        $name = '')
    {
      $this->code = $code;
      $this->name = $name;
    }
  }
```

Finally, we can run `SchoolClassTest` successfully. Now, we have to adjust the files of the user interface, which are stored in the `view/schoolclass` folder. Why do we need to make this adjustment? For two reasons. The first is that now we will use the `getAll` method from the object returned by `SchoolClassTable:getInstance`. The second is that elements of returned collections are not arrays but objects.

Adjusting CRUD files

Let's start with the file that lists the school classes:

list.php

```php
<?php
use School\SchoolClassTable;

require realpath(__DIR__ . '/../../vendor') .'
    /autoload.php';

$baseUrl = '/view/schoolclass/';

foreach(SchoolClassTable::getInstance()->getAll() as $row):
?>
<tr>
    <td><a href="<?=$baseUrl?>edit.php?code=<?=$row->
        code?>"><?=$row->code?></a></td>
    <td><?=$row->name?></td>
    <td><a href="<?=$baseUrl?>delete.php?code=<?=$row->
        code?>">Remove</a></td>
</tr>
<?php
endforeach;
```

Observe that the $row variable is an object. If you remember, in the equivalent file of the school1 project, the $row variable was an array. Now, the records of table classes are mapped to objects of the SchoolClass class.

The file with the edit form must also be adapted to use SchoolClassTable and the object attributes instead of array elements:

edit.php

```php
<?php
use School\SchoolClassTable;
require realpath(__DIR__ . '/../../vendor') .'
    /autoload.php';
$code = $_GET['code'] ?? null;
$schoolClass = SchoolClassTable::getInstance()->
    getByField('code',$code);
?>
<!DOCTYPE html>
<html>
    <head>
     <meta charset="UTF-8">
     <title>Classes Registration</title>
     </head>
     <body>
     <h1>Classes Registration</h1>
     <form action="save.php" method="post">
     Name: <input type="text" name="name" autofocus=
     "autofocus" value="<?=$schoolClass->name?>"><br/>
    <input type="hidden" name="code" value="<?=
        $schoolClass->code?>"><br/>
    <input type="submit" value="save">
    </form>
    <p>
    <a href="index.php">Go back</a>
    </p>
    </body>
</html>
```

The edit.php file submits the data to save.php, like its counterpart in the school1 project. The save.php file makes the persistence with two classes, SchoolClass and SchoolClassTable. The first class has the data and the second class has the operations.

The save.php file will look like this:

save.php

```php
<?php
use School\SchoolClass;
use School\SchoolClassTable;
require realpath(__DIR__ . '/../../vendor') .
    '/autoload.php';
$code = (int)($_POST['code'] ?? 0);
$name = ($_POST['name'] ?? '');
$model = new SchoolClass($code,$name);
$schoolClassTable = SchoolClassTable::getInstance();
$schoolClassTable->save($model);
header('Location: /view/schoolclass/index.php');
```

Objects must help us as developers. They are tools, not commandments. When it is helpful to use an object, use it. But if there is no reason to use an object, avoid it. Following this guideline, we don't need to create an object to remove a school class record.

It is enough to pass the code, as we can see in the delete.php file:

delete.php

```php
<?php
use School\SchoolClassTable;
    require realpath(__DIR__ . '/../../vendor') .'
        /autoload.php';
$code = ($_GET['code'] ?? null);
$schoolClassTable = SchoolClassTable::getInstance();
$schoolClassTable->delete($code);
header('Location: /view/schoolclass/index.php');
```

Now, you can manage school classes through the web interface of school2. But we promised that this project would also implement student registration. The implementation of both registrations (school classes and students) is similar, so we will only show the parts that are different.

We will have a `StudentTable` class that extends `AbstractTable`. But unlike `SchoolClassTable`, the `StudentTable` class overrides the `getByField` and `getAll` methods. This is because the default query of `TableGateway`, made with the `select` method, reads only from a table, ignoring relationships. To add an `INNER JOIN` clause to `SQL SELECT` executed by `TableGateway`, we need to use an instance of the `Select` class:

```
public function getAll(): iterable
{
    $select = new Select($this->tableGateway->getTable());
    $select->join('classes', 'classes.code =
        students.class_code', [
      'class_name' => 'name'
    ]);
    $rowSet = $this->tableGateway->selectWith($select);
    return $rowSet;
}
```

The `Select` class has a `join` method, which allows setting an `INNER JOIN` clause. The third parameter of the `join` method allows us to select the columns of the table and specify an alias as a key of the array. To use the query with `INNER JOIN`, we inject the `Select` object into the `selectWith` method of `TableGateway`. The `selectWith` method returns a `ResultSet` object, like the `select` method.

The `Student` class also has differences when we compare it with the `SchoolClass` class. The `Student` class overrides the `exchangeArray` and `toArray` methods, inherited from `AbstractModel`.

The `exchangeArray` method is overridden because it needs to instantiate the `SchoolClass` class. One student depends on a school class. This is a constraint of our use case. We cannot have students without a school class to associate them with. As we can see in the implementation of `Student->exchangeArray`, the `Student` class has a `$this->schoolClass` attribute:

```
public function exchangeArray($data)
{
    $this->id = ((int) $data['ID'] ?? 0);
    $this->name = ($data['name'] ?? '');
    $this->schoolClass = new SchoolClass();
    $this->schoolClass->code = ((int) $data['class_code']
        ?? 0);
    $this->schoolClass->name = ($data['class_name'] ?? '');
}
```

The class_name field is the alias for the field name of the classes table. We have defined this alias in INNER JOIN made through the Select class.

Although we handle the data in memory as objects, we need to convert the data into arrays to persist them in the table. So, in the Student->toArray method, we replace the object with the primary key value:

```php
public function toArray()
{
    $attributes = get_object_vars($this);
    $attributes['class_code'] = $attributes['schoolClass']
        ->code;
    unset($attributes['schoolClass']);
    return $attributes;
}
```

To insert a new student, we need to associate it with a school class. This requires a list of school classes in the edition form for students. In the next part of the code, we can see the part of the edit.php file of view/student that creates this list:

```php
<select name="class_code">
<?php
$schoolClasses = SchoolClassTable::getInstance()->getAll();
foreach($schoolClasses as $schoolClass):
?>
<option value="<?=$schoolClass->code?>"
    <?=($schoolClass->code == $student->schoolClass->code ?
    'selected="selected"' : '')?>>
     <?=$schoolClass->name?></option>
<?php
endforeach;
?>
</select>
```

These are the main differences between school classes and student registrations. The school2 project is fully available at https://github.com/PacktPublishing/PHP-Web-Development-with-Laminas/tree/main/chapter04/school2. There is a StudentTest test case for student registration and a report of the code coverage. Try the complete application and compare the code of both registrations. In Eclipse, you can compare two files by selecting them and pressing the right mouse button, following which you select **Compare with/Each Other**.

Now you know how to abstract a database connection and how to create ORM using `laminas-db`.

In the next section, we will make a new implementation of the school classes and student registrations, but this time we will use the `laminas-mvc` component as a container for our application.

Using Laminas as a container

When you use Laminas as a library, your code calls the Laminas components. It's what we did with the `school1` and `school2` projects. Now, we will use Laminas as a framework and this framework will call your code. Previously, Laminas was contained in our applications. Now, we will create an application that will be contained in the framework. In fact, we will rewrite our code according to the MVC structure of Laminas.

As this is a preparation for a more complex and complete application, we will show only the implementation for the listing of the school classes, but do not worry, because the complete code is available at the end of the section. But try to implement the code along with the explanation before copying it from the repository, so you can compare your expectations with the provided code.

Like we did before, we need to prepare for the implementation by creating a new project. Let us create it!

Creating the school3 project

To create the `school3` project, follow these steps:

1. Create the project.

   Do you remember that we have configured an item in the Eclipse toolbar that creates Laminas projects? We did it in the previous chapter. Well, you can use it to create the `school3` project.

   Open the project using the **File | New | PHP Project** menu option.

2. Add **PHPUnit** to the `school3` project as a development dependency.

3. Don't forget to save the `composer.json` file and click on the **Update dependencies** button.

> **Allow plugins**
>
> When you are installing a component after creating the project from the skeleton, you might be asked **Do you trust 'laminas/laminas-component-installer' to execute code and wish to enable it now?** and **Do you trust 'laminas/laminas-skeleton-installer' to execute code and wish to enable it now?**. These messages are related to Composer plugins. You can answer y.

4. After installing PHPUnit, select the `school3` project with the right mouse button and access the **Run As | PHPUnit Test** item. You will see the following message on the **Console** tab:

```
Fatal error: Uncaught Error: Class "Laminas\Test\PHPUnit\
Controller\AbstractHttpControllerTestCase" not found
in /opt/lampp/htdocs/school3/module/Application/test/
Controller/IndexControllerTest.php:11
```

This error occurs because the Laminas skeleton project depends on the `laminas-test` component, an extension for PHPUnit that allows us to test the Laminas controllers. As you have installed PHPUnit in the `composer.json` editor, search for the `laminas/laminas-test` package and install it. We learned how to install a Composer package in *Chapter 3, Using Laminas as a Library with Test-Driven Development*.

5. Now you can run the test successfully. But what test is running? The `IndexControllerTest` class in the `module/Application/test/Controller` folder.

> **Errors when testing Laminas MVC in PHP 7**
>
> Maybe you remember that in *Chapter 2, Setting Up the Environment for Our E-Commerce Application*, we showed PHP requirements for `laminas-mvc`. The `composer.json` file allows PHP 7. However, if you run `IndexControllerTest` under PHP 7, there will be an error for the `testIndexActionViewModelTemplateRenderedWithinLayout` test method. In fact, the error occurs because the `$this->assertQuery` method depends on union types, available only from PHP 8. You can replace `$this->assertQuery` with other assertion methods and keep using `laminas-test` under PHP 7, but it is better to upgrade PHP to version 8. But if you are required to use PHP 7 for some reason, then remove the `testIndexActionViewModelTemplateRenderedWithinLayout` method.

Now that we have run the test, we will create a new module.

Creating the School module

The Laminas MVC application is organized as a group of modules. A module in Laminas is the biggest unit of reuse. A module is self-contained. It has its own configuration and its own MVC implementation. To implement our school class registration with `laminas-mvc`, we will create another module, named `School`. The Laminas modules are located in the `modules` folder. There is a default module named `Application`. The `Application` module contains the sample home page that you viewed in the previous chapter.

We will use the `Application` module as the model for a new module:

1. Copy the `Application` folder in the `modules` folder as `School` in the same `modules` folder.

The namespace of a module in Laminas is the name of the module. So, the namespace of the `Application` module is `Application` and, therefore, the namespace of the `School` module must be `School`. We will change the namespace of each file as we rebuild our implementation. Let's start with the minimal code for the module.

2. Register the module in the `modules.config.php` file.

    The `modules.config.php` file is in the `config` folder in the project root. This file contains the list of enabled modules as a PHP array. Modules outside of this list are ignored, as if they don't exist. This makes it easy to disable a module by removing it from `modules.config.php`. You can see that `Application` is the last item. It is not random. The last module is what imposes the layout. The layout is the arrangement of visual elements. For a web application, the layout is defined with **Cascade Style Sheets (CSS)**. Our new module, `School`, does not have a layout, so it can't be the last.

3. Add `School` as an item before `Application`.

    A module listed in `modules.config.php` will be searched for. But the modules will only be found if they have autoloading configurations. You should remember that we have configured autoloading via Eclipse for the `school1` project. You can use the same feature to configure `composer.json` of `school3`.

4. Select the **Autoload** tab in the Composer code editor. You will see that there is already a PSR-4 configuration for the `Application` module. You must add a new autoloading configuration with these parameters:

    -   **Namespace**: `School\`
    -   **Paths**: `module/School/src/`

    We have to add a PSR-4 autoloading configuration for test classes too.

5. Select the **Autoload-dev** tab. Add a configuration with these parameters:

    -   **Namespace**: `SchoolTest\`
    -   **Paths**: `module/School/test/`

6. Save the `composer.json` file and click on the **Update dependencies (including require-dev)** button.

Editing composer.json directly

Every change we are making with specific forms can also be made directly. You can click on the **composer.json** tab at the bottom of the composer.json view. The editor will open the file as a text file. You can see the changes made by other tabs and you can also make changes directly.

Now the `School` module can be found. But it will not work yet. We must ensure that each file inside the `School` module has the right namespace.

Ensuring the module has the right namespace

The process of ensuring the module has the right namespace involves certain steps. Let us start as follows:

1. Create a test class. As we copied the `Application` module as a `School` module, we have an `IndexControllerTest.php` file inside the `module/School/test/Controller` folder. Delete this file and create a new file named `SchoolClassControllerTest.php` in the same folder.

   This new file must contain a `SchoolClassControllerTest` class that extends `AbstractHttpControllerTestCase`, which is part of the `laminas-test` component. It provides assertion methods to test the control layer of `laminas-mvc`. This is something that we lacked in our previous projects. You should remember that we tested the model classes, but the requests depended on our interaction with the web browser. Now we can cover this part without using a web browser or making HTTP requests manually.

   The namespace of the `SchoolClassControllerTest` class is `SchoolTest`, as we have configured in `composer.json`.

2. Implement a `setUp` method to load the application configuration:

```
public function setUp(): void
{
    $this->setApplicationConfig(
        include __DIR__ . '/..//../../config/
            application.config.php'
    );

    parent::setUp();
}
```

This `setUp` method creates an object that acts as a container for the application. We will learn about it in detail in the next chapter. For now, assume that you will have access to all resources of the application that you need for testing.

Implementing the test methods

Now, we can start implementing the test methods using the following steps:

1. Create a method to test the school classes listing, `testListing`:

    ```
    public function testListing(): void
    {
        $this->dispatch('/schoolclass', 'GET');
        $this->assertResponseStatusCode(200);
        $this->assertCommonRules();
    }
    ```

    The `testListing` method emulates an HTTP request for endpoint `/schoolclass` and expects an HTTP response with the status 200.

    Right, but what about the call for the `$this->assertCommonRules` method? Well, we have advanced here. The fact is that the test methods of `SchoolClassControllerTest` have common assertions because they make requests to the same controller class.

2. Instead of repeating the same lines, we encapsulate them into a private method:

    ```
    private function assertCommonRules(): void
    {
        $this->assertModuleName('school');
        $this->assertControllerName('schoolclass');
        // as specified in router's controller name alias
        $this->assertControllerClass
            ('SchoolClassController');
        $this->assertMatchedRouteName('schoolclass');
    }
    ```

    The `assertCommonRules` method has the following assertions:

    -   `$this->assertModuleName`: Checks the module – the argument value is converted following the CamelCase practice. What does this mean? The value `school` will be converted to `School`, with the first letter being in uppercase, which is the real module name.

    -   `$this->assertControllerName`: Checks the alias for the controller name – this will be seen ahead.

    -   `$this->assertControllerClass`: Checks the name of the controller class (without the namespace).

    -   `$this->assertMatchedRouteName`: Checks whether the pattern of the URL request matched a configured route.

3. The Laminas application skeleton, which we have used since *Chapter 2, Setting Up the Environment for Our E-Commerce Application*, provides a template file for PHPUnit, `phpunit.xml.dist`. Copy it to `phpunit.xml`. You can see that the test suite is configured for running only the tests of the `Application` module:

```
<testsuite name="Laminas MVC Skeleton Test Suite">
   <directory>./module/Application/test</directory>
</testsuite>
```

4. Change the preceding code snippet to run the tests for any module:

```
<testsuite name="Laminas MVC Skeleton Test Suite">
   <directory>./module/*/test</directory>
</testsuite>
```

Great! We can take the first step of TDD, the red step, when we have a test that fails.

Running the tests

Follow these steps to run the tests:

1. Select the `school3` project and click on the **Run As | PHPUnit Unit Test** item.

   Of course, the test will fail because `SchoolClassController` doesn't exist.

2. Create `SchoolClassController` to manage the generation of the school classes listing. The file of this class must be created in the `module/SchoolClass/src/Controller` folder. In fact, all the classes of a module in Laminas stay in the `src` folder and all the controller classes stay in the `src/Controller` folder. This class needs two methods: a constructor in which we inject the Table Data Gateway and a method for invoking the table data gateway:

SchoolClassController.php

```php
<?php
declare(strict_types = 1);
namespace School\Controller;

use School\Model\SchoolClassTable;
use Laminas\Mvc\Controller\AbstractActionController;
use Laminas\View\Model\ViewModel;


class SchoolClassController extends
```

```
    AbstractActionController
{

  private ?SchoolClassTable $schoolClassTable = null;

  public function __construct(SchoolClassTable
      $schoolClassTable)
  {
    $this->schoolClassTable = $schoolClassTable;
  }

  public function indexAction()
  {
    $schoolClasses = $this->schoolClassTable->
        getAll();
    return new ViewModel([
      'schoolClasses' => $schoolClasses
    ]);
  }
}
```

As we can see, the constructor method receives an instance of `SchoolClassTable` and assigns it to the `$this->schoolClassTable` attribute. The `indexAction` method uses this attribute to get the list of school classes and deliver it to the view layer through the `ViewModel` class.

If you don't need PHP 7 compatibility, you can locate the following:

```
private ?SchoolClassTable $schoolClassTable = null;

public function __construct(SchoolClassTable
    $schoolClassTable)
{
  $this->schoolClassTable = $schoolClassTable;
}
```

Replace it with this:

```
public function __construct(private SchoolClassTable
    $schoolClassTable)
```

```
    {
    }
```

This is possible from PHP 8.

3. Create a `Model` folder. Change the namespaces in the target to `School\Model`.

   Right, but where does the `SchoolClassTable` class come from? We already created it!

4. Copy all the files from the `src` folder of the `school2` project to the `modules/School/src/Model` folder.

   Now, most of the code for the model layer is ready. However, we need to change some lines. Let's return to the controller class to understand this.

   As you saw in the last source code, the `SchoolClassController` class receives an instance of `SchoolClassTable`. But who injects this instance? Not you, because the controllers in an MVC Laminas application are not directly instantiated. You need to use a factory class.

5. Create a factory class named `SchoolClassControllerFactory` in the `src/Controller` folder of the `School` module:

SchoolClassControllerFactory.php

```php
<?php
namespace School\Controller;

use Interop\Container\ContainerInterface;
use Laminas\ServiceManager\Factory\FactoryInterface;

class SchoolClassControllerFactory implements
    FactoryInterface
{
    public function __invoke(ContainerInterface
    $container, $requestedName, ?array $options = null)
    {
        $schoolClassTable = $container->get
            ('SchoolClassTable');
        return new SchoolClassController
            ($schoolClassTable);
    }
}
```

The `SchoolClassControllerFactory` class uses a dependency injection container to recover the instance of `SchoolClassTable`. In fact, the get method of the `$container` object uses a key to locate the factory for `SchoolClassTable`. This key takes us to a class named `SchoolClassTableFactory`.

6. Create a `SchoolClassTableFactory` class using the following source code:

SchoolClassTableFactory.php

```php
<?php
declare(strict_types = 1);
namespace School\Model;

use Laminas\Db\TableGateway\TableGateway;
use Laminas\Db\TableGateway\TableGatewayInterface;
use Laminas\Db\ResultSet\ResultSet;
use Laminas\ServiceManager\Factory\FactoryInterface;
use Interop\Container\ContainerInterface;

class SchoolClassTableFactory implements FactoryInterface
{
  public function __invoke(ContainerInterface
  $container, $requestedName, ?array $options = null)
  {
    $adapter = $container->get('DbAdapter');
    $resultSetPrototype = new ResultSet();
    $resultSetPrototype->setArrayObjectPrototype(new
        SchoolClass());
    $tableGateway = new TableGateway('classes',
        $adapter, null, $resultSetPrototype);
    return new SchoolClassTable($tableGateway);
  }
}
```

You might say: "I already saw this somewhere else..." You are right. This code is the content of the static `getInstance` method of `SchoolClassTable`. Thus, `SchoolClassTable` doesn't need this method anymore because the responsibility to create instances was given to the factory class.

7. Remove the `getInstance` method of `SchoolClassTable` and change the constructor of this class to public. This means that you will need to locate the following line:

```
private function __construct(TableGatewayInterface
$tableGateway)
```

Replace it with this line:

```
public function __construct(TableGatewayInterface
$tableGateway)
```

The controller factory class depends on a specific configuration to find the data table gateway factory class.

The key used by the `get` method of the `$container` object is an alias. This alias must be defined in a configuration file. As we are using `SchoolClassTable` in the `School` module, we will define this configuration in the `module.config.php` file of the `School` module. This file is in the `config` folder of the module, which makes sense.

8. Change the namespace of `module.config.php` to `School`. This file is in fact a PHP array with the module configuration. You will find a key named `controllers`, as follows:

```
'controllers' => [
  'factories' => [
    Controller\IndexController::class =>
        InvokableFactory::class,
  ],
],
```

This code snippet shows that the `IndexController` class is associated with the `InvokableFactory` class. `InvokableFactory` is a controller factory provided by `laminas-mvc`. This basic factory creates an instance of the associated class without dependency injection. We don't use `IndexController`.

9. Remove the `IndexController.php` file from the `src/Controller` folder. The previous code snippet must be changed, as follows:

```
'controllers' => [
  'aliases' => [
    'schoolclass' => Controller\
        SchoolClassController::class
  ],
  'factories' => [
    Controller\SchoolClassController::class =>
```

```
            Controller\SchoolClassControllerFactory::class
        ],
    ],
```

Yes, besides changing the association in factories, we have also added key `aliases` associating the `"schoolclass"` string with `SchoolClassController`. What does the association do? It helps in decoupling the route from the real controller names.

10. Change the `routes` key, inside the `router` key. Delete the current value of `router` and replace it with the following:

```
'routes' => [
  'schoolclass' => [
    'type'  => Segment::class,
    'options' => [
      'route'  => '/schoolclass[/:action
          [/:code]]',
      'defaults' => [
        'controller' => 'schoolclass',
        'action'    => 'index',
      ],
    ],
  ],
],
```

A route is an association between a URL pattern and a controller class. There are several route types available in the `laminas-router` component. In this case, we are using a `Segment` route that allows parameters. In the `'/schoolclass[/:action[/:code]]'` pattern, the part between brackets is optional and the word after the colon is a param. The `defaults` key defines assumed values for omitted params. See that we defined default values for `controller` and `action`. The value of the `controller` key is the alias for `SchoolClassController`.

You may have forgotten, but the value of an alias is the value passed in the `$this->assertControllerName` method of `SchoolClassControllerTest`. In fact, you can use the alias or the complete name of the class.

Can we run the test case successfully now? Well, the test will still fail because we need to configure the alias for `SchoolClassTable`.

11. Configure the alias by adding the `service_manager` key to the array of `module.config.php`. This key is at the first level of the following array:

```
'service_manager' => [
  'factories' => [
    'SchoolClassTable' => Model\
      SchoolClassTableFactory::class
  ]
],
```

The dependency injection container will use the keys defined inside factories to call the factories declared as values.

Now can we run the test? No.

12. Change the namespace of the `Module` class in the `School/src` folder to `School`. The `Module` class loads the `module.config.php` file, so if it is not found, the module configuration is ignored.

After this change, we can run the test, right? No, because we need to configure the database connection. Keep calm and configure. We will keep using the `school` database, the same database used for the `school1` and `school2` projects.

13. Install `laminas-db`. Fortunately, you know how to do it. But this time, as you are installing `laminas-db` inside a project with `laminas-mvc`, you will see on the **Console** tab the following prompt:

```
Please select which config file you wish to inject
'Laminas\Db' into:
[0] Do not inject
[1] config/modules.config.php
[2] config/development.config.php.dist
Make your selection (default is 1)
```

This allows you to inject (or not) `laminas-db` as a module for the application. As we have mentioned before, a module in Laminas is the biggest unit of reuse. You can inject a component only for development (option 2). But we will inject `laminas-db` as a component for production, so choose option 1.

14. Composer will next ask the following:

```
Remember this option for other packages of the same type?
(Y/n)
```

Answer Y, so you avoid the previous prompt for the next Laminas components to be installed.

Remember that laminas-db needs connection parameters. According to Laminas' best practices, we must separate the global configuration (valid for any environment) from the local configuration (valid for a specific environment).

15. Place the global configuration in the global.php file in the config/autoload folder in the project root. The content for the school project must be as follows:

global.php

```php
<?php

use Laminas\Db\Adapter\AdapterServiceFactory;
return [
  'db' => [
    'driver'   => 'Pdo_Mysql',
    'database' => 'school'
  ],
  'service_manager' => [
    'factories' => [
      'DbAdapter' => AdapterServiceFactory::class
    ]
  ]
];
```

The difference between the database configuration and previous projects is the use of AdapterServiceFactory, nicknamed DbAdapter. This class uses the data of the db key to create an instance of our old friend: the Adapter class. Right, but what about the other parameters? Well, the other parameters can be different depending on the environment. So, we must put them into the local.php file. While global.php is the same for all the environments, there is a different local.php for each environment. Because of this, the local.php file must not be part of the control version. The skeleton brings a template named local.php.dist.

16. Copy `local.php.dist` from the `config/autoload` folder to `local.php` in the same folder and fill it with this code:

---

local.php

```php
<?php
return [
    'db' => [
        'hostname' => 'localhost',
        'username' => 'root',
        'password' => ''
    ]
];
```

We have all the necessary data. You must be excited to see whether the test runs successfully now.

17. Run the test:

```
1) SchoolTest\Controller\
SchoolClassControllerTest::testListing
Failed asserting response code "200", actual status code
is "500"

Exceptions raised:
Exception 'Laminas\View\Exception\RuntimeException' with
message 'Laminas\View\Renderer\PhpRenderer::render:
Unable to render template "school/school-class/index";
resolver could not resolve to a file' in /opt/lampp/
htdocs/school3/vendor/laminas/laminas-view/src/Renderer/
PhpRenderer.php:492
```

When you run, you receive the preceding error message. This message occurs because there is no view for the controller action.

Observe that the error message says clearly that the application expects a `school/school-class/index` template. There is a convention between the Laminas controller and view layers. The controller methods that receive the HTTP requests from the routes have an `Action` suffix. When an `Action` method ends and returns a `ViewModel` object, the framework will search for a file with the same name as the method, but without the `Action` suffix. So, the `indexAction` method is related to a file named `index`. The default extension for view files is `.phtml`.

18. Create an `index.phtml` file in the `view/school/school-class` folder inside the School module.

   The `view` folder has an `application` subfolder, a copy of the folder from the Application module.

19. Remove this subfolder because it is not useful for the School module.

20. Create a `school` folder inside the `view` folder and a `school-class` folder in the school folder.

   The first level of folders under `view` is related to controller classes. The second level is related to controller actions. The name of folders and files inside the `view` folder must be in small letters. When the name of a class or method is formed of several names, the related folders and filenames must be separated with a hyphen.

21. Create the `index.phtml` file inside the `school-class` folder, as follows:

index.phtml

```
<h1>Classes Registration</h1>
<p>
  <a href="<?=$this->url('schoolclass',
         ['action' => 'edit'])?>">Add a new class</a>
</p>
<table>
  <thead>
    <tr>
      <th>code</th>
      <th>name</th>
    </tr>
  </thead>
  <tbody>
<?php
    foreach($this->schoolClasses as $row):
?>
      <tr>
        <td><a href="<?=$this->url
          ('schoolclass',['action' => 'edit', 'code'
          => $row->id])?>"><?=$row->code?></a></td>
```

```
            <td><?=$row->name?></td>
            <td><a href="<?=$this->url('schoolclass',
                    ['action' => 'delete', 'code' => $row-
                        >code])?>">Remove</a></td>
        </tr>
    <?php
        endforeach;
    ?>
        </tbody>
        </table>
    <p><a href="<?=$this->url('home')?>">Homepage</a></p>
```

The `$this->schoolClasses` attribute was defined by the `indexAction` method of `SchoolClassController` through the `ViewModel` class. The true difference for this file is the use of the `$this->url` method. This method creates a URL from a route declared in some of the `module.config.php` files.

22. Finally, you can run the test successfully!

    Wait, won't we need to navigate in the web interface? Well, you can navigate partially with the listing implemented here or by downloading the complete project.

23. Start a web server. As we did with our "hello world" project, you can use the embedded web server of PHP. Inside the `school3` folder, you run the following:

    php -S localhost:8000 -t public

    Now you can access the `school3` application at `localhost:8000`.

    Don't worry if some details here seem like magic; each one will be explained starting in the next chapter. The main aim of *Chapter 3, Using Laminas as a Library with Test-Driven Development and Chapter 4, From Object-Relational Mapping to MVC Containers* was to avoid a sudden leap from an experience with pure PHP to the Laminas way of development. You can find a quick-start guide in Laminas' documentation, but it is like an improved "hello world," without database handling.

So, you have reached the end of this section with a little MVC Laminas application, where the framework calls your code instead of your code calling the framework components.

The `school3` project is fully available at `https://github.com/PacktPublishing/PHP-Web-Development-with-Laminas/tree/main/chapter04/school3`. You can install, navigate, and handle school classes and students. After this preparation, you are ready for our awesome e-commerce project. You will soon deeply understand each layer of the Laminas MVC implementation.

Summary

In this chapter, we have learned how to use `laminas-db` to implement ORM.

There are ORM generation tools for PHP, such as **Doctrine**, that create classes for you from the database schema. However, code generation tools can sometimes produce unnecessary code, as well as taking control of ORM from the developer. With `laminas-db`, the developer has total control over ORM and can avoid superfluous code that code generation tools sometimes produce.

We gradually evolved the code base until we could use `laminas-mvc` to contain our code. We also kept best practices in mind to create the tests before creating the code to be tested. This experience with TDD will prepare us for the next challenges.

In the next chapter, we will increase the complexity level of web application development. We will start the implementation of an e-commerce project that allows us to explore many features provided by Laminas components.

Part 2: Creating an E-Commerce Application

In this part, you will obtain the necessary knowledge to build a web application for e-commerce with the best practices of secure development. By the end of this part, you will have a complete working web application using the main components of Laminas.

This section comprises the following chapters:

- *Chapter 5, Creating the Virtual Store Project*
- *Chapter 6, Models and Object-Relational Mappers with Behavior-Driven Development*
- *Chapter 7, Request Control and Data View*
- *Chapter 8, Creating Forms and Implementing Filters and Validators*
- *Chapter 9, Event-Driven Authentication*
- *Chapter 10, Event-Driven Authorization*

5

Creating the
Virtual Store Project

This chapter presents the requirements of the e-commerce project we plan to build, its use cases, its class diagrams, and the creation of the general structure of the application. We will start a more complex application that can serve as a reference for creating PHP enterprise web applications.

By the end of this chapter, you will be able to create the structure for a complex software project using the Laminas framework as a container and service provider.

In this chapter, we'll be covering the following topics:

- Understanding the project requirements for the sample web application

- Identifying use cases and actors

- Understanding the class diagram for *whatstore* (our fictional department store)

- Creating the project instance

Technical requirements

The environment for the virtual store project has been prepared in *Chapter 2, Setting Up the Environment for Our E-Commerce Application*. The hardware reference configuration has been presented in *Chapter 3, Using Laminas as a Library with Test-Driven Development*. The code generated for this chapter is available in full at `https://github.com/PacktPublishing/PHP-Web-Development-with-Laminas/tree/main/chapter05`.

Understanding the project requirements for the sample web application

We will imagine a fictional backstory for our project. You are a software company owner. Your company is called *Sheat Software*. One day, your company is hired to create an e-commerce website for a department store called *Whatstore*. You may think that the name of the department store rather sounds like a question, but in fact, the name is an acronym for *We Have All Things Store*. This company really sells many products. Unfortunately, the contract doesn't include suggesting a better name for the client company. So, we can instead focus on the software construction.

After some meetings with Mr. Heythere, CEO of Whatstore, some requirements were defined for the virtual store:

1. The virtual store system must have two distinct zones: one for the customers (sales area) and another for the employees (management area). The first area must allow customers to put products into a basket and pay for them. The second area must allow employees to register products, make the inventory, change prices, and give discounts.

2. The home page of the customer area must allow the searching of products by name, completely or partially.

3. The product pages must show the names of products, their prices, and the number in stock. For each product, there must be a button to add the item to the basket.

4. There must be an always-visible link for the basket that allows the customer to see all the selected products, change the amount of each one, remove products, or close the order.

5. The customers should be able to register themselves at any point in the process. It should not be necessary for them to register themselves while selecting products, but while checking out, the customer should automatically be registered.

6. When the order is closed, the requested number of products must be reserved in the inventory. When the products are prepared for delivery, the number must be subtracted from the inventory.

7. The system must have access control for the management area that allows them to define who can do what.

These are the functional requirements that provide a general view of the requested system. In the next section, we will refine these requirements for use cases. As we have experienced in the previous chapter, a use case can help us develop, not only because it provides the flows but also because we can produce a test case from a use case. A test case, in turn, helps us to maintain the stability of the software.

Identifying use cases and actors

In the previous chapter, we made a simplified use case specification (managing school classes). In this chapter, we will present here only the use case diagrams. For our purposes, the diagrams and the subsequent comments will be enough.

There are two main actors for the *Whatstore* virtual store: the customer and the employee. First, we will present the use case diagram for the *employee* actor. Before the customer buys something from the store, the store needs to have products. *Figure 5.1* shows the use cases related to the maintenance of the product inventory:

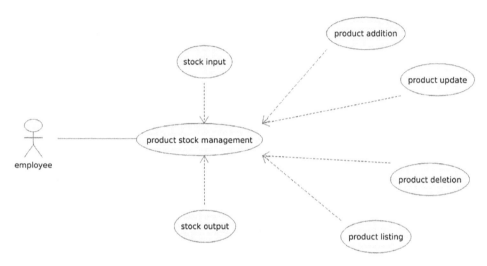

Figure 5.1 – Use cases for the employee actor

As we can see, our e-commerce system will have a product registration. Well, you already have the experience of creating registrations after the examples in the previous chapter. But the product inventory has some operations beyond **Create, Read, Update, and Delete** (**CRUD**). The amount of a certain product that the store has is stored in a different table of the product registration, as we saw in *Chapter 2, Setting Up the Environment for Our E-Commerce Application*. The amount of each product available can be changed through specific operations related to the acquisition or sale of items.

You may find it strange that *Figure 5.1* presents some use cases that would only be operation flows according to the approach that we have used in *Chapter 3, Using Laminas as a Library with Test-Driven Development*. You must remember that the CRUD operations in the school system were part of the same use case. This was convenient for our preparation because we needed to have a more simplified view of the project. So, it was better to group the flows in just one use case. From now, we will deal with more details, so it is better to divide the operations to create smaller and more readable tests.

Since we have available products for sale, the customer can interact with the system. As we can see in *Figure 5.2*, there are three main flows for the customer: registration, product selection, and order closure:

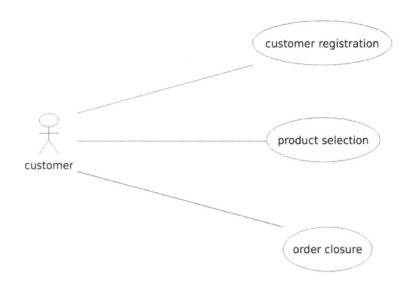

Figure 5.2 – Use cases for the customer actor

Customers can register themselves. But what about employees? Who adds an employee to the system? We can see that there should be an actor who creates the employees: a manager. A manager can also be an employee, but one who plays a management role. *Figure 5.3* shows the use cases associated with the *manager* actor:

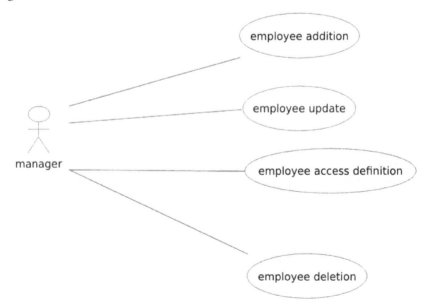

Figure 5.3 – Use cases for the manager actor

So, we have identified three actors for our system: *customer*, *employee*, and *manager*. The actors of use cases can help us identify classes, although this doesn't mean that every actor will be turned into a class.

In *Chapter 2*, *Setting Up the Environment for Our E-Commerce Application*, we have presented the **model-entity relationship (MER)** for our application. In fact, this artifact is part of the general requirements definition. The entities also help us to identify classes. We have worked, in the last chapter, with object-relational mapping, and we have seen that there is a similarity between classes and tables.

From the combination of use cases and MER, next, we will present the class diagram, which will be used for the implementation of the models for the application.

Understanding the class diagram for Whatstore

Figure 5.4 shows the class diagram for the Whatstore system. You can observe that there are two sets of classes: a set of independent classes, Role and Employee, and another set of classes that has dependencies. The coupling among classes of the use cases related to the customer should be clear, considering that most of the requirements mention the customer searching, selecting, or ordering products.

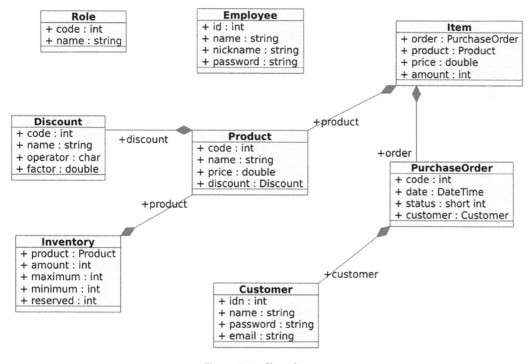

Figure 5.4 – Class diagram

Where did the Role class come from? Previously, we said that the manager is a role played by an employee. The role of a manager allows an employee to insert, update, and delete other employees from the system. This role can be associated with more than one employee if the CEO wishes it. The current managers can lose their role at any time. We could define a `Manager` class for dealing with this specific scenario of privileges. This requires further thought.

What if a new role was requested for the system? For example, a role that can insert employees, but can't update them or remove them. Should we create a new class? We try to avoid unnecessary changes in a software system. Each change usually adds more complexity, and this tends to make it more difficult, over time, to add new features and find the cause of bugs.

We have to allow changes in a software system in a way that don't require changes in the source code. The architecture must be flexible to allow adjustments via configuration. We have to search for structures that allow behavior changes that occur due to the inserting of new data or from the association between elements of the system. If we have a generic structure for roles, we can create a new role at any time. The behavior of the employees can be changed by the association between employees and roles.

The software will change. Let's focus on the changes that add value to the system (and to our budget). Think about the future and try to create generic structures that can be easily adapted to specific cases. This desirable feature of flexibility is the reason for the decoupling between the `Role` and `Employee` classes.

Of course, we are trying to simplify an e-commerce system with our fictional story of *Whatstore*. We will try to cover as many of the main features as possible and will take every opportunity to comment on architectural aspects that can be useful for any complex web application project.

For the moment, we have the necessary inputs. In the next section, we will start the implementation with the main architectural structures based on the Laminas framework.

Creating the project instance

How wonderful is it to start on a new project after having worked on something else previously? You are more confident now. You have a prepared environment with suitable tools. Let's practice what we have learned in the previous chapters. Using the **Create Laminas Project** external tool configuration in Eclipse, create a project named `whatstore`. Remember that, after creating, you need to open the project, using the **File | New | PHP Project** menu. In fact, Eclipse won't create a new PHP project, but it will open an existing project in the filesystem.

After creating the project, you can start the application using the embedded PHP web server (considering that you are running inside the `whatstore` directory) as follows:

```
php -S localhost:8000 -t public
```

If you open the `localhost:8000` address in the browser, you will see the welcome page of Laminas, as you have done previously in *Chapter 2, Setting Up the Environment for Our E-Commerce Application*. But why do we use the `-t` argument with the `public` value? We will understand this in the next section.

Understanding the structure of a Laminas project

Let's understand the structure of a Laminas project from the directories of a new project:

- `bin`: This folder is used for storing scripts for automation
- `config`: This folder stores the application configuration
- `data`: This folder is used for storing data that doesn't need to be shared, such as cache files
- `module`: This is the heart of the application, the folder that contains the modules, which we talk more about in this section
- `public`: This is the public part of the application, the only folder that can be accessible to the users
- `vendor`: This is the default folder for dependencies managed by Composer

So, we start the application by pointing to the `public` folder because `public` is the web root for the users. It's the only point of access for the users, avoiding the unnecessary exposition of the main part of the application. This separation between `public` and the rest makes the application akin to a castle surrounded by a deep moat (which can have crocodiles or worse). The `public` folder is like the drawbridge of this castle. You must walk over the bridge to enter the castle. So, in the configuration of the filesystem for the web server, you can deny access to any folder except `public`.

We have said that the `module` folder is the heart of the application. This is true. Try to remove this folder and see what happens. Well, the `module` folder can have one or more modules, which are in a subfolder with a name according to a CamelCase pattern. Let's see the content of the default `Application` module to understand the general structure of a module:

- `config`: This folder stores the module configuration
- `src`: This folder stores the classes that implement the use cases
- `test`: This folder stores the test cases
- `view`: This folder stores template files for the HTML pages

This structure is well suited to an MVC implementation. The elements of the model and controller layers are in the `src` folder. The elements of the view layer are in the `view` folder. This is the proposed and default structure for a Laminas module. It is possible to change everything in this structure, but there is no need to, in general. Laminas has its patterns, but it is configurable.

The real configuration for runtime is the combination of the application configuration with the modules' configuration. The application configuration deals with shared resources, such as database connections used by several modules. A module configuration, at first, must deal only with the resources contained by the module. However, a module can use resources from another module, although it is not recommended to make high couplings.

Laminas' MVC implementation has a `ModuleManager` class that orchestrates the modules as the only application. `ModuleManager` is responsible for merging several modules' configurations into a single configuration. *Figure 5.5* shows a simplified scheme of the configuration reading of a module, a process that occurs inside a loop:

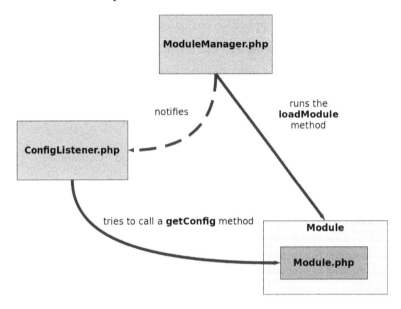

Figure 5.5 – Configuration reading by a listener

In fact, there are more classes and interfaces involved in this process, but we will see more about that in *Chapter 7, Request Control and Data View*. For this moment, it is enough to know the main elements.

Each module has a `Module` class inside the `src` folder. `ModuleManager` searches for the modules listed in the `modules.config` file (in the application configuration). A `ConfigListener` class is notified when `ModuleManager` loads a module. `ConfigListener` tries to call a `getConfig` method from the `Module` class. The `getConfig` method, if it exists, must return a configuration array. The module configuration contains, as we have seen in the previous chapter, the routes for the module, the controllers that can be instantiated, the source for the view templates, and the factories for any object that needs to be created with dependency injection.

In the next section, we will prepare the modules for our project.

Preparing the modules for our project

We will have three modules for the *Whatstore* application. We will keep the `Application` module and we will create two more: `Inventory` and `Store`. We will use the `Application` module as the source for the other modules. Copy the `Application` directory inside the module folder and paste it two times, one as `Inventory` and the other as `Store`.

You can observe that the mere existence of the Inventory and Store folders in the folder module doesn't affect the application, by loading the welcome page at localhost:8000. ModuleManager will ignore these folders until their names are in the modules.config.php file in the config folder. Add the names of the new modules in this file as follows:

module.config.php

```
return [
    'Laminas\Router',
    'Laminas\Validator',
    'Store',
    'Inventory',
    'Application',
];
```

The Application module must stay at the end of this list because it will define the layout of the web pages. We will talk about this in *Chapter 7, Request Control and Data View*.

After adding the new modules to the configuration, we will see this error message in the web server log:

```
PHP Fatal error:  Uncaught Laminas\ModuleManager\Exception\
RuntimeException: Module (Store) could not be initialized.
```

What is the problem? ModuleManager doesn't find the modules. But how? They are in the modules folder! Wait, calm down! We already covered this in *Chapter 4, From Object-Relational Mapping to MVC Containers*. If you need a refresher, then reread the *Using Laminas as a container* section.

We have said that Laminas is configurable. It doesn't force you to follow a rigid structure. You have the flexibility to adjust the project structure. So, you can put a module in a different place and there is a reason for this. This is possible because the loading of classes is delegated to Composer. We define the autoloading rules in the composer.json file. You need to change the autoload and autoload-dev keys as follows:

```
"autoload" : {
  "psr-4" : {
    "Application\\" : "module/Application/src/",
    "Inventory\\" : "module/Inventory/src/",
    "Store\\" : "module/Store/src/"
  }
},
"autoload-dev" : {
```

```
    "psr-4" : {
      "ApplicationTest\\" : "module/Application/test/",
      "InventoryTest\\" : "module/Inventory/test/",
      "StoreTest\\" : "module/Store/test/"
    }
  },
```

After finishing the autoloading configuration, don't forget to save the `composer.json` file and click on the **Update dependencies (including require-dev)** button in the Composer editor in Eclipse. This action will regenerate the `autoload` files.

But when you watch the output in the **Console** tab, you will see that this action also searches for updates, which makes sense. How can we regenerate the `autoload` files without a search for updates? We can create an external tool configuration, as we did in the *Integrating Laminas and Eclipse* section in *Chapter 2, Setting Up the Environment for Our E-Commerce Application*, for Laminas project creation. You can create a configuration like this:

- **Name**: Composer `autoload` files regeneration
- **Location**: Here, you must type the complete path to the Composer executable
- **Working Directory**: `${project_loc}`
- **Arguments**: `dumpautoload`

So, when you need to regenerate the `autoload` files, you run this configuration by the Eclipse toolbar.

However, even regenerating the `autoload` files, the modules are not loaded yet. Why? We have seen this before. Didn't we copy the `Application` module? Then, all the namespaces of the `Inventory` and `Store` modules are `Application` too. Then you need to edit all the classes and configuration files in these modules and change the namespace for the namespaces defined in the `composer.json` file. For example, for the `Inventory` module, the `modules.config.php` file and the classes in the `src` folder must have the `Inventory` namespace. The classes in the `test` folder, in their turn, must have the `InventoryTest` namespace.

You can check that, after fixing the namespaces, you can successfully load the welcome page of the Laminas skeleton for the *Whatstore* application. It seems that all is right for the modules. You can install PHPUnit and `laminas-test` as development dependencies for running the tests and the result will be positive, but this will be a false positive because we didn't create the `phpunit.xml` file, including all the test cases. Copy the `phpunit.xml.dist` template as `phpunit.xml`. Find the following line:

```
<directory>./module/Application/test</directory>
```

And change it to this line:

```
<directory>./module/*/test</directory>
```

This may feel like dèjá vú because you already did this in *Chapter 4, From Object-Relational Mapping to MVC Containers*, (at least I hope you did). Now, if you run `phpunit` again, you will see errors for the two new modules. These errors are related to the routes called by the tests. But for now, we will not fix the test classes. We will do so in *Chapter 7, Request Control and Data View*.

As the routes for the new modules are not configured yet, the new controllers are not instantiated and therefore, the controllers didn't try to render the template views. But it is important to talk once more about the structure of the `view` folder for preparing our modules for the next implementations.

The `view` folder of a Laminas module must be a folder with the name of the module, but only with lowercase letters. The names of folders and files under the `view` folder are all in lowercase letters. Then, change the `application` folder under the `view` folder of the `Inventory` and `Store` modules to `inventory` and `store`, respectively. Remove the `layout` folder under the `view` folder of these modules too. Only one module needs to have a `layout` folder and the `Application` module already has one. The appearance of the new modules should be as shown in *Figure 5.6*:

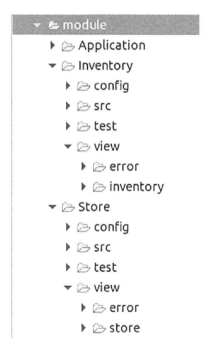

Figure 5.6 – Structure of the folder module

This is sufficient to start a new Laminas project with custom modules. In the next chapter, we will implement the model layer of our application, using the projected classes in this chapter.

Summary

In this chapter, we have presented the requirements of our main sample web application: an e-commerce system for a fictional company, *Whatstore*. We have strived to understand the business rules and requirements of their requested system. We have used some use case diagrams to show at a high level the main flows of the application. In addition, we have presented the projected classes that will serve as inputs for the model layer implementation.

We have also reviewed some steps done in *Chapter 2, Setting Up the Environment for Our E-Commerce Application*, and *Chapter 4, From Object-Relational Mapping to MVC Containers*, for the creation of a project with Composer (mediated by Eclipse) and the creation of a new module, using the `Application` module as a template.

In this chapter, you have learned how to start a complex software project with the Laminas framework. We can understand complexity as a concept related to the number of things in a system and the difficulty in knowing and managing each one of them. According to Cartesian principles, to solve a complex problem, we need to break it into smaller parts. Here, we have defined a modular structure for our software project, dividing the main problem into two parts (two modules). In subsequent chapters, we will learn about division inside modules, which will help us to focus on one small problem at a time.

Contemporary software is complex, and it is necessary to have approaches and tools to deal with complexity. The modular structure of the Laminas framework combined with MVC patterns helps PHP web developers manage complex applications for a distributed systems world.

In the next chapter, we will review what we already learned about the `laminas-db` component, and we will delve into the ORM implementation. We will deal with code reuse when we have to create a generic implementation for our mapper classes.

Models and Object-Relational Mappers with Behavior-Driven Development

In this chapter, we will use `laminas-db` to map an entity-relationship model to an object model. Instead of just implementing mappers with Laminas, we will also create a most powerful reusable mapper extending Laminas classes. Finally, we will show how models and mappers are connected. Along with the creation of the classes, we will use the approach of **behavior-driven development (BDD)** to keep user-readable requirements in our source code.

By the end of this chapter, you will have learned how to create models and object-relational mapper classes using the approach of BDD. You will know how to map system requirements to code in a well-organized way. I hope you finish the chapter feeling as satisfied as Luke Skywalker did after exploding the Death Star!

In this chapter, we'll be covering the following topics:

- BDD with Behat
- Creating models from user stories
- Creating mappers from user stories
- Creating a generic model and a generic mapper
- Creating other models with Behat

Technical requirements

All the user stories implemented for **Behat** for this chapter are fully available at https://github.com/PacktPublishing/PHP-Web-Development-with-Laminas/tree/main/chapter06/whatstore/features.

In the same repository, you can find the model classes in the `Model` folder of the `Inventory` and `Store` modules.

BDD with Behat

In *Chapter 3, Using Laminas as a Library with Test-Driven Development*, we experienced the TDD approach. TDD is a very important skill to ensure the stability of software. When you practice TDD, you know that all the software continues to work after you change a part of it. Beyond this, the test classes document the business rules, and thus, TDD helps you to maintain useful documentation. A test class with running tests tells you that the rules contained in the tests are enforced by classes or subroutines of the software.

Test cases written in the xUnit pattern could be more readable for non-programmers than the direct implementation of the software, but you are still using a programming language to express business rules. How wonderful would it be if we could write tests in a more human language? BDD provides this. It is an approach that extends TDD, using an abstraction layer over the automated tests, written by and for programmers. It allows the writing of more understandable stories for those who know a business but don't know, or even need to know, the technology they are using to develop their applications.

> **The xUnit pattern**
>
> Kent Beck and Erich Gamma created JUnit, a test framework for Java. The JUnit architecture was used as a base for creating test frameworks for other languages such as PHP. There is a family of test frameworks that follow the same pattern for the implementation of unit tests, test cases, and test suites. Kent Beck and other authors such as Gerard Meszaros call this family *xUnit*, where the letter *x* is a variable placeholding the programming language. The common structure of xUnit frameworks is well described by Gerard Meszaros in his book *xUnit Test Patterns: Refactoring Test Code* (Addison-Wesley).

We will use BDD to implement the model layer for the `Whatstore` project. We need to install a BDD tool for PHP. There is a BDD implementation for PHP called *Behat*.

Using the Composer editor of Eclipse, add a development dependency for the `behat/behat` package. Remember that you already did something like this in *Chapter 3, Using Laminas as a Library with Test-Driven Development*, when you installed PHPUnit. After adding the package, save the file, and click on the **Update dependencies (including require-dev)** button to download Behat.

Based on what we have learned in *Chapter 2, Setting Up the Environment for Our E-Commerce Application*, we will create some external tool configurations for the most common invocations of the Behat executable. Using the **Run** | **External Tools** | **External Tools Configurations** menu, create the following three configurations:

- **Behat initialization**:

  Name: `Behat init`

Location: `${project_loc}/vendor/bin/behat`

Working Directory: `${project_loc}`

Arguments: `--init`

- **Running of the behavior tests**:

  Name: `Behat run`

  Location: `${project_loc}/vendor/bin/behat`

  Working Directory: `${project_loc}`

  Arguments: Leave this empty

- **Snippet appending**:

  Name: `Behat append snippets`

  Location: `${project_loc}/vendor/bin/behat`

  Working Directory: `${project_loc}`

  Arguments: `--append-snippets`

Don't forget, as we did in *Chapter 2, Setting Up the Environment for Our E-Commerce Application*, to check the **External Tools** item on the **Common** tab for each one of these configurations.

Now that we have avoided unnecessary command typing for the next steps, we can initialize Behat for the `Whatstore` project. Select the project name in the **Project Explorer** view and click on the **Behat init** item on the external tools list in the toolbar, or in the **Run | External Tools** menu. The output on the **Console** tab will be as follows:

```
+d features - place your *.feature files here
+d features/bootstrap - place your context classes here
+f features/bootstrap/FeatureContext.php - place your
definitions, transformations and hooks here
```

> **Variable references empty selection: $(project_loc)**
>
> If the `Variable references empty selection: $(project_loc)` message appears after you run any external tool, it is because you haven't selected the project name in the **Project Explorer** view. Select the `Whatstore` project name and run it again.

Press *F5* or click on the *Refresh* item of the project to view the new `features` folder in the **Project Explorer** view. The files with the `.feature` extension contain user stories, with a language closer to the human language than a programming language. We can say that it seems like a *normal people language*. The `bootstrap` folder will contain classes with code related to the stories defined by the `.feature` files. Don't worry – we will explain how to create stories.

Select the project name in the **Project Explorer** view and click on the **Behat run** item on the external tools list in the toolbar, or in the **Run/External Tools** menu. The output on the **Console** tab will be this:

```
No scenarios
No steps
0m0.00s (8.67Mb)
```

There are no scenarios because there are no files with the .feature extension inside the features folder. Let's create our first scenario to understand what it is in BDD.

Creating our first scenario

In the preceding chapter, we presented several use cases. Some of them were related to the management of products by employees. Let's start with the use case *product addition*. A use case is, in fact, a user story that expresses a desired feature for a software system. So, from the use case *product addition*, we will write a file named productaddition.feature that will describe this feature. Create this file through the **File** | **New** | **File** menu. Its content will be the following:

---

productaddition.feature

```
Feature: Product addition
  In order to have products in the stock
  As an employee
  I need to be able to add a product to the product
  registration

  Rules:
  - The code of product is generated automatically
  - Product name has a maximum of 80 characters
  - A product must have a discount, and the default
    discount is the null discount

  Scenario: Inserting a product in the registration
    Given there is a product called "Troop Power Battery",
    which costs $2814
    When I add this product to the registration
    Then I should have a product called "Troop Power
    Battery" in the table products
    And the overall product price should be $2814
```

Let's understand the patterns used for the `productaddition.feature` file. The file begins with the word `Feature`. A feature is another name for a user story or use case. The three lines that follow have initial words that must be present for every feature description:

- `In order...`: The goal of the user story
- `As a/an...`: The role played by the story actor
- `I need...`: The condition for the actor to reach the goal

These initial words define a pattern for describing a feature. Optionally, you can add a `Rules` section to describe the business rules related to the feature. This first part of the `.feature` file is only descriptive and works like any other document of requirements. The second part, on the other hand, serves to generate source code.

A scenario describes a flow using an example. So, we have real values instead of variables. You can create several blocks, starting with the word `Scenario`, according to the possible flows. A `Scenario` block has phrases that start with the following words:

- `Given...`: The condition for the flow to start
- `When...`: The action that starts the flow
- `Then...`: The result of the flow
- `And` (optional)...: An additional result of the flow

All these four keywords are part of a language projected to give structure and meaning to executable specifications – **Gherkin**. Gherkin is part of the common language analyzer of **Cucumber**, a BDD tool for *Java*. You can learn more about Gherkin at this URL: `https://cucumber.io/docs/gherkin/reference`.

The phrases of a scenario are used to create methods of a **context class**. A context class allows you to map a user story to an automated test. Let's see how it works.

Mapping a user story to an automated test

As you have seen, we wrote a scenario where an employee inserts a product called `"Troop Power Battery"` that costs $2,814. As the type of discount was not specified, we assume that the discount is `null`. Okay, but how does this scenario become an automated test? You will discover this by selecting the `whatstore` project in the **Project Explorer** view and clicking on the **Behat run** item in the **Run | External Tools** menu. The output on the **Console** tab will have the content of the `productaddition.feature` file added with the following lines:

```
1 scenario (1 undefined)
4 steps (4 undefined)
0m0.01s (8.89Mb)
```

```
>> <snippet_undefined><snippet_keyword>default</snippet_
keyword> suite has undefined steps. Please choose the context
to generate snippets:</snippet_undefined>

  [0] None
  [1] FeatureContext

>
```

The preceding output says that there is an undefined scenario. Why? Because there is no context class with code that implements the scenario. In the end, Behat waits for you to select a target to code snippets for the scenario. You can select None, and nothing will be done, or you can select the default context class, FeatureContext. For now, choose None because we will create a specific class for our scenario.

Now, copy the FeatureContext.php file from the features/bootstrap folder to the ProductAdditionContext.php file in the same folder. Change the name of the class inside the file to ProductAdditionContext. Is this procedure enough for Behat to show this new class as a target option? No, because the default behavior of Behat is to search only for the FeatureContext class. If you don't believe me, try to run Behat now. You will see that ProductAdditionContext will be ignored.

To change the behavior of Behat, we must create a configuration file. Create a file named behat. yml in the features folder with the following content:

behat.yml

```
default:
    autoload:
        '': '%paths.base%/bootstrap'
    suites:
        default:
            paths:
                - '%paths.base%'
            contexts:
                - ProductAdditionContext
```

Behat supports multiple configuration profiles. We will use only the default profile. The autoload key defines where Behat will find the context classes. The suites key allows the definition of multiple sets of context classes suites. In this case, we only have a default set of context suites. The paths key defines where the user story files will be searched.

We need to change the **Behat run** external tool, specifically the **Arguments** field. Replace the current content with the following code:

```
--config ${project_loc}/features/behat.yml
```

Now, you can run the Behat external tool and the output will be as follows:

```
[0] None
[1] ProductAdditionContext
```

You don't need to remove `FeatureContext.php`. In fact, this file is a template. You can use it directly for learning, but this class must not be used for real contexts, except when you have a user story called *feature*, which is very strange. Well, let's come back to the **Console** tab. The console is waiting for your choice – 0 or 1? Type 1, which is `ProductAdditionContext`. The result will be as follows:

```
/**
 * @Given there is a product called :arg1, which costs
       $:arg2
 */
public function thereIsAProductCalledWhichCosts
    ($arg1, $arg2)
{
    throw new PendingException();
}


/**
 * @When I add this product to the registration
 */
public function iAddThisProductToTheRegistration($arg1)
{
    throw new PendingException();
}


/**
 * @Then I should have a product called :arg1 in the
       table products
 */
public function iShouldHaveProductInTheTableProducts
```

```
        ($arg1)
    {
        throw new PendingException();
    }


    /**
     * @Then the overall product price should be $:arg1
     */
    public function theOverallProductPriceShouldBe($arg1)
    {
        throw new PendingException();
    }
```

Observe that the generated source code has methods whose names are the phrases of the `Inserting a product in the registration` scenario. Observe also that the values in the file were replaced in the PHP source code with variables such as `$arg1`. We will see how this is useful going forward. However, if you were waiting to see the generated code snippets inserted into the `ProductAdditionContext` class, you would discover that this code is not there.

Yes, the **Behat run** external tool only shows the source code generated from the user story. To add this code to the context class, we need to run the **Behat append snippets** external tool. This external tool also needs to use the customized configuration, so change the value of the **Arguments** field this way:

Arguments: `--append-snippets --config ${project_loc}/features/behat.yml`

What are you waiting for? Run it! This command also asks which context you want. Type 1 for `ProductAdditionContext`, and you will see the following output:

```
u bootstrap/ProductAdditionContext.php - `there is a product
called "Troop Power Battery", which costs $2814` definition
added
u bootstrap/ProductAdditionContext.php - `I add the "Troop
Power Battery" to the registration` definition added
u bootstrap/ProductAdditionContext.php - `I should have a
product called "Troop Power Battery" in the table products`
definition added
u bootstrap/ProductAdditionContext.php - `the overall product
price should be $2814` definition added
```

You can open the `ProductAdditionContext.php` file and check that snippets were inserted into it. Does it mean that we can run the behavior test with **Behat run** now? No. If you run **Behat run** now, you will see the following output:

```
Scenario: Inserting a product in the
registration                            # ProductAddition.
feature:11
    Given there is a product called "Troop
Power Battery", which costs $2814 #
ProductAdditionContext::thereIsAProductCalledWhichCosts()
        TODO: write pending definition
    When I add this product to the
registration                      #
ProductAdditionContext::iAddThisProductToTheRegistration()
    Then I should have a product called "Troop Power
Battery" in the table products                      #
ProductAdditionContext::iShouldHaveAProductCalledInTheTable
Products()
    And the overall product price should
be $2814                           #
ProductAdditionContext::theOverallProductPriceShouldBe()


1 scenario (1 pending)
4 steps (1 pending, 3 skipped)
0m0.01s (9.31Mb)
```

Now, the phrases of the scenario have, as comments, the name of the correspondent methods in the context class. In the end, we can see that scenario is not undefined anymore but pending. The steps have changed to 1 pending and 3 skipped. The state pending means that the context class has methods to implement the steps of the scenario; however, these methods have no content yet. It's like a PHPUnit test case failing because there is no code to be tested yet.

We have to complete the scenario implementation. For this, we need to implement the models and object-relational mappers related to the scenario. We will start this in the next section.

Creating models from user stories

In a grammar class, you may have heard the question, "*In this phrase, who or what is the subject?*" In a similar way, we ask the following: "*In an* Inserting a product in the registration *scenario, what is the model?*" Yes, exactly – it is the product! Then, we need to instantiate a Product class in our behavior test class. For this, we must define a $product attribute in the ProductAdditionContext class:

```
private ?Product $product = null;
```

We can instantiate the Product class in the constructor of the ProductAdditionContext class and use this instance in the thereIsAproductCalledWhichCosts method, as follows:

```php
public function __construct()
{
    $this->product = new Product();
}


/**
  * @Given there is a product called :arg1, which costs
     $:arg2
  */
public function thereIsAProductCalledWhichCosts
    ($arg1, $arg2)
{
    $data = [
        'name'  => $arg1,
        'price' => $arg2
    ];
    $this->product->exchangeArray($data);
}
```

Observe that we pass $arg1 (product name) and $arg2 (product price) as arguments to the exchangeArray method of the Product class. Of course, if you try to run **Behat run** now, a failure will occur because the Product class doesn't exist yet. We are in the first step of TDD – a test that fails. Remember that BDD is an extension of TDD, and then we keep the TDD steps in BDD. If the test fails, we are in debt. We need to write the code for the test run successfully. Let's create the Product model.

> **Note – Behat arguments in definitions**
>
> The values identified as arguments in the feature files are mapped to context classes with the aid of the @Given, @When, and @Then annotations. These annotations inform whether the following method has or does not have an argument. If you change the value of an argument in a definition, you don't need to change the implementation. For example, if the product price is changed in the productaddition.feature file, you don't need to change anything in the ProductAdditionContext class. This avoids the need to make changes in classes if only a value from a rule has changed.

The Product class will stay inside the Inventory module. Create a Model folder in the src folder of this module to contain the model classes. The source code of the Product model will be as follows:

Product.php

```php
<?php
namespace Inventory\Model;

class Product
{
    public int $code;
    public String $name;
    public float $price;
    public Discount $discount;

    public function __construct()
    {
        $this->exchangeArray([]);
    }

    public function exchangeArray($data): void
    {
        $this->code = ($data['code'] ?? 0);
        $this->name = ($data['name'] ?? '');
        $this->price = ($data['price'] ?? 0.0);
        $this->discount = new Discount();
        $this->discount->code = ($data['code_discount'] ??
            0);
    }
}
```

Why did you implement a constructor calling the `exchangeArray` method? To initialize the attributes. Otherwise, we can suddenly get an `attribute must not be accessed before initialization` error.

As you can see, the `Product` class depends on the `Discount` class. Thus, we have to create the `Discount` class as follows:

Discount.php

```php
<?php
namespace Inventory\Model;
```

```php
class Discount
{
    public int $code;
    public string $name;
    public string $operator;
    public float $factor;

    public function __construct()
    {
        $this->exchangeArray([]);
    }

    public function exchangeArray($data): void
    {
        $this->code = ($data['code'] ?? 0);
        $this->name = ($data['name'] ?? '');
        $this->operator = ($data['operator'] ?? '');
        $this->factor = ($data['factor'] ?? 0.0);
    }
}
```

Maybe another framework would have generated the model classes from the table schemas. Perhaps. But to be generic, such a framework could have generated unnecessary code too. We are creating only the necessary code for our models.

Well, after creating the models, you can return to the `ProductAdditionContext` class and add the namespace:

```php
use Inventory\Model\Product;
```

Then, you can execute **Behat run** again, with a different summary result:

```
1 scenario (1 pending)
4 steps (1 passed, 1 pending, 2 skipped)
0m0.09s (9.31Mb)
```

Now, the first step has passed the behavior test. You can observe that this step is printed in green.

The next step is to add the product to the registration. For this, we need to create a table mapper. This is the subject of the next section. Don't worry about the other models; we will come back to them later.

Creating mappers from user stories

Before writing new code in the `ProductAdditionContext` class, we will reuse some code written in the training projects. If you remember the `school3` project, we created global and local configurations for the database connection. Well, you can copy the `global.php` and `local.php` files from the `school3` project to the `whatstore` project. These two files stay in the `config/autoload` folder. After copying them, you need to rename the database in the `global.php` file to `whatstore`.

We also need to install `laminas-db` to implement object-relational mapping. We did this in *Chapter 4, From Object-Relational Mapping to MVC Containers*. Then, use Composer to install `laminas-db`, as we did once. Don't forget that is not enough to add the dependency to the `composer.json` file. You must run `composer update` in the Composer Eclipse editor by clicking the **Update dependencies** button. The following question should appear on the console:

```
Please select which config file you wish to inject 'Laminas\Db'
into:
```

Choose the option that injects into the `config/modules.config.php` file. Why? Read *Chapter 4, From Object-Relational Mapping to MVC Containers*, again. Next, another question should be done:

```
Remember the option for other packages of the same type?
(Y/n)?:
```

Answer Y to avoid the repetition of this question.

We touched on object-relational mapping with `laminas-db` in *Chapter 4, From Object-Relational Mapping to MVC Containers*, section *Implementing the test methods*. You may remember that we work with an implementation of the **Table Data Gateway** pattern. This pattern is implemented in Laminas by the `TableGateway` class. There are some repetitive steps to use this class for specific mapper classes. For this, we created some abstractions in the `school3` project. It's time to reuse them. Copy the `AbstractModel.php` and `AbstractTable.php` files from the `School` module in the `school3` project to the `Inventory` module in the `whatstore` project. These files are part of the model layer, so they must stay in the `src/Model` folder. Change the namespace of these classes in the destination for `Inventory\Model`. Ready! You will avoid repetitive code. Let's code the addition of the product to the registration.

Adding a product to the products table

We will change the constructor of the `ProductAdditionContext` class to instantiate a **Table Data Gateway** implementation, as follows:

```
public function __construct()
{
    $this->product = new Product();
    $this->productTable = $this->getApplication()->
        getServiceManager()->get('ProductTable');
}
```

As we can see, we are setting an `$this->productTable` attribute. Therefore, this attribute must be declared at the beginning of the class:

```
private ?ProductTable $productTable = null;
```

The `$this->productTable` attribute receives an object from the `get` method of the `Laminas\ServiceManager\ServiceManager` class. We recover the instance of `ServiceManager` from an instance of `Laminas\Mvc\Application`. This last instance is created by a private `getApplication` method that we must implement in the following way:

```
private function getApplication()
{
    $appConfig = require __DIR__ . '/../../config/
        application.config.php';
    if (file_exists(__DIR__ . '/../../config/
        development.config.php')) {
        $appConfig = ArrayUtils::merge($appConfig,
            require __DIR__ . '/../../config/
                development.config.php');
    }

    return Application::init($appConfig);
}
```

The `Application` instance is created by the `init` method of this class, from the application configuration.

Since we have a table mapper, we can insert a product. The `ProductAdditionContext->iAddThisProductToTheRegistration` method will call the `save` method of the `ProductTable` class this way:

```
public function iAddThisProductToTheRegistration()
{
    $this->productTable->save($this->product);
}
```

Right, but where does the `save` method come from? Well, the `ProductTable` class must be created through the following implementation:

ProductTable.php

```php
<?php
declare(strict_types = 1);
namespace Inventory\Model;

use Laminas\Db\TableGateway\TableGateway;
use Laminas\Db\TableGateway\TableGatewayInterface;
use Laminas\Db\ResultSet\ResultSet;

class ProductTable extends AbstractTable
{

    protected string $keyName = 'code';

    protected string $modelName = 'Inventory\Model\
        Product';

    public function __construct(TableGatewayInterface
        $tableGateway)
    {
        $this->tableGateway = $tableGateway;
    }
}
```

As you can see, the `ProductTable` class extends the `AbstractTable` class. The save method is implemented by `AbstractTable`, so we only need to inform which model class (the `$modelName` attribute) will be handled in the subclass.

Does it mean a product can be inserted now? In fact, if you execute **Behat run**, you will see the following error message:

```
Unable to resolve service "ProductTable" to a factory; are you
certain you provided it during configuration?
```

Why does this happen? It is because we need to define a factory for `ProductTable`. The `ServiceManager` class does not create direct instances with the new operator. `ServiceManager` searches for a factory from an alias. We are using the name of the table mappers as an alias, so the factory alias for the `ProductTable` class is `ProductTable`. But how is a real factory class? Well, a factory for a data mapper stays in the same `Model` folder of the model class handled by this mapper. So, we have to create the following class in the `Model` folder of the `Inventory` module:

ProductTableFactory.php

```php
<?php
declare(strict_types = 1);
namespace Inventory\Model;

use Interop\Container\ContainerInterface;
use Laminas\Db\ResultSet\ResultSet;
use Laminas\Db\TableGateway\TableGateway;
use Laminas\ServiceManager\Factory\FactoryInterface;

class ProductTableFactory implements FactoryInterface
{
    public function __invoke(ContainerInterface $container,
        $requestedName, ?array $options = null)
    {
        $adapter = $container->get('DbAdapter');
        $resultSetPrototype = new ResultSet();
        $resultSetPrototype->setArrayObjectPrototype
            (new Product());
        $tableGateway = new TableGateway('products',
            $adapter, null, $resultSetPrototype);
```

```
            return new ProductTable($tableGateway);
    }
}
```

The `ProductTableFactory` class is very similar to factories that we created in the `school3` project in *Chapter 4, From Object-Relational Mapping to MVC Containers*. The `__invoke` method of this class encapsulates the dependency chain to create an instance of `ProductTable`. Since we have created the factory, we will configure an alias for the `ServiceManager` instance that searches for this factory. `ServiceManager` uses aliases to invoke object `factories` because it is a fast way to access a value in a map configuration using a key. We will make this in the `module.config.php` file of the `Inventory` module. We will add a key at the end of the array, like this:

```
    'service_manager' => [
        'factories' => [
            'ProductTable' => Model\ProductTableFactory::
                class
        ]
    ]
```

And now? Will the product be inserted? Not yet. If you execute **Behat run**, you will see the following error message:

```
Type error: Inventory\Model\AbstractTable::save(): Argument
#1 ($model) must be of type Inventory\Model\AbstractModel,
Inventory\Model\Product given
```

This happens because our `Product` model class is not a subclass of `AbstractModel`. But this is easy to solve. Add `extends AbstractModel` in the declaration of the `Product` class.

After this change, you can execute **Behat run**. Then, you will see the following error:

```
Type error: PDO::quote(): Argument #1 ($string) must be of type
string, Inventory\Model\Discount given
```

The preceding error occurs because the `toArray` method of `AbstractTable` returns all the attributes of `Product`, and one of them is an instance of the `Discount` class. However, the `ProductTable->save` method needs scalar values to mount a SQL INSERT. So, we have to override the `toArray` method in the `Product` class in the following way:

```
    public function toArray()
    {
        $attributes = get_object_vars($this);
```

```
unset($attributes['discount']);
$attributes['code_discount'] = $this->discount->
    code;
return $attributes;
}
```

Observe that we remove the object attribute and add only the value for the foreign key. Finally, we can execute **Behat run** with a successful output:

```
1 scenario (1 pending)
4 steps (2 passed, 1 pending, 1 skipped)
0m0.23s (12.66Mb)
```

Great! We have implemented two steps of the scenario: the `thereIsAProductCalledWhichCosts` and `iAddThisProductToTheRegistration` methods of the `ProductAdditionContext` class. Let's now check whether the product was inserted.

Checking whether the product was inserted

As we didn't reach the end of our scenario, we are not reverting the changes in the database yet. Manually remove the record inserted in the last execution of **Behat run** to avoid trash in the `products` table.

Now, we will implement the assertions to check whether the product was really inserted into the `products` table. These assertions must be implemented into the methods of the `ProductAdditionContext` class, named according to the expected results – that is, into the `iShouldHaveProductInTheTableProducts` and `theOverallProductPriceShouldBe` methods. The source code is as follows:

```
/**
 * @Then I should have a product called :arg1 in the
   table products
 */
public function iShouldHaveAProductCalledInTheTable
    Products($arg1)
{
    $this->product = $this->productTable->getByField
        ('name',$arg1);
    Assert::assertEquals($arg1, $this->product->name);
}
/**
```

```
    * @Then the overall product price should be $:arg1
    */
   public function theOverallProductPriceShouldBe($arg1)
   {
        $product = $this->productTable->getByField
            ('name',$this->product->name);


        Assert::assertEquals($this->product->name,
            $product->name);
   }
```

Observe that we are using assertion methods of the `Assert` class, which belongs to PHPUnit. This means that we need to declare the namespace of this class:

```
use PHPUnit\Framework\Assert;
```

Finally, we see that Behat integrates with PHPUnit. You already learned how to develop from test cases. Now, you have learned how to develop from user stories. If you execute **Behat run** now, the output will be the following:

```
1 scenario (1 passed)
4 steps (4 passed)
0m1.49s (13.62Mb)
```

> **Note – Behat does not include PHPUnit**
>
> Behat does not have PHPUnit as a dependency by default. Thus, to use the `Assert` class or any other class of PHPUnit in Behat context classes, you need to install PHPUnit separately. In our case, we have already installed PHPUnit.

As we have seen, the scenario is completely tested. But wait! After the `ProductAdditionContext` behavior test class is run, we have to create a record in the `products` table. As this insertion was done by a test, we have to remove the record to return to the previous state. Otherwise, each test will increase the table by one record.

First, manually remove the record that you have inserted. Then, insert the following method into the `ProductAdditionContext` class:

```
   public function __destruct()
   {
        $product = $this->productTable->getByField
            ('name',$this->product->name);
```

```
        $this->productTable->delete($product->code);
}
```

This is a destructor method. This method is executed automatically before the instance is destroyed. We recover the inserted product from the table and delete it. Now, if you execute **Behat run**, you will notice that there is no new record in the `products` table. This is because the record is deleted at the end of the behavior test – that is, you can execute the behavior test as many times as you need. The number of records in the `products` table after the test will be the same as before the test.

We have now learned how to implement a table mapper from user stories. In the next section, we will return to the models, explaining how we will deal with the generalization of the model layer.

Creating a generic model and a generic mapper

We have actually already created a generic model (the `AbstractModel` class) and a generic mapper (the `AbstractTable` class). We have used these generalizations in this chapter, respectively, as the superclass of `Product` and the superclass of `ProductTable`. Really, it seems that the generic model and generic mapper are ready now. But these generalizations are not in the appropriate place. If you look, the `AbstractModel` and `AbstractTable` classes are in the `Inventory` module. The problem is that the `Inventory` module is not the only module to have models. We also have the `Store` module. We will use the generic model and mapper in both of the modules. For this, we will extract the generic model and mapper to a `Generic` module. Follow these steps:

1. Copy the `Application` module to `Generic`. This means copying the `Application` folder to the `Generic` folder.

2. Remove the `config`, `test`, and `view` folders from the `Generic` module.

3. Remove the `IndexController.php` file from the `module/Generic/src/Controller` folder. This file was copied from the `Application` module, but it is not necessary for the `Generic` module.

4. Change the namespace of the `Module` class in the `src` folder to `Generic` and remove the `getConfig` method from this class.

The `Generic` module will work as a library for the other modules. We will put any abstract class or interface shared by more than one module of our application into the `Generic` module.

You may have noticed that the constructor methods of table mappers are exactly the same. Therefore, let's improve the code for reuse by moving the constructor to the `AbstractTable` class. Cut the following part of the `ProductTable` class:

```
public function __construct(TableGatewayInterface
        $tableGateway)
    {
```

```
            $this->tableGateway = $tableGateway;
    }
```

Paste the preceding code into the `AbstractTable` class. From now on, the table mappers won't need to declare the constructor.

> **Note – To inherit or to use? That is the question**
>
> Our table mapper abstract class (`AbstractTable`) uses an implementation of `TableGatewayInterface`, from Laminas. This approach allows us to change easily the **Table Data Gateway** pattern implementation by any other implementation. The dependency injection provides flexibility and vendor independence. However, if you accept establishing a strong coupling with Laminas oriented-object mapping, there is the option to inherit the `TableGateway` or `AbstractTableGateway` classes. To inherit can be more convenient, but, as we have said, using classes through dependency injection provides more flexibility. To use is better than to be, when you consider that changes in software are a certainty.

As we did with previous modules, it is necessary to configure autoload rules for this new module. Edit the `composer.json` file and add the following line to the `autoload/psr-4` key:

```
"Generic\\": "module/Generic/src"
```

In Eclipse, you can make this change by clicking on the **composer.json** tab of the Composer editor, as shown in *Figure 6.1*:

Figure 6.1 – The composer.json file with the autoload rule for the Generic module

You may ask, why don't we add an autoload rule for `Generic` tests, as we have done for the `Inventory` and `Store` modules? We don't because it is not necessary. If you test the classes that extend classes of the `Generic` module, both the `Inventory` and `Store` modules will be covered. In fact, we can't test effectively abstract classes or interfaces because they don't have implementations – at least not all the implementations.

Save the `composer.json` file and execute the **Composer autoload files regeneration** external tool. This item is available on the toolbar, as *Figure 6.2* shows:

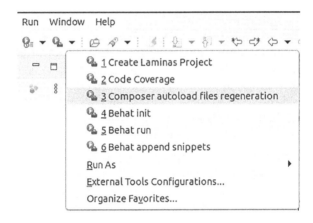

Figure 6.2 – The Composer autoload files regeneration external tool

This external tool, **Composer autoload files regeneration**, was created in *Chapter 5*, *Creating the Virtual Store Project*. You can see that we have configured several external tools in previous chapters. Make a note that we created **Composer autoload files regeneration** because we need to recreate the composer autoload files to reflect the content of `composer.json`. Is it enough for the `Generic` module to be found? No. Although Laminas can find the classes in the `Generic` namespace, the `ModuleManager` class will search only for the modules listed in the `modules.config.php` file. Add an entry for the `Generic` module so that the `modules.config.php` file looks like this:

modules.config.php

```php
<?php
return [
    'Laminas\Db',
    'Laminas\Router',
    'Laminas\Validator',
    'Generic',
    'Store',
```

```
    'Inventory',
    'Application',
];
```

Now, we can refactor the `whatstore` project, moving the generic model and mapper from the `Inventory` module to the `Generic` module. Let's go through the following steps:

1. Create a `Model` folder in `modules/Generic/src`.

2. Select the `AbstractModel.php` and `AbstractTable.php` files in `modules/Inventory/src/Model`.

3. Move these files to `modules/Generic/src`.

4. Change the namespace of both of the files to `Generic\Model`.

After preparing the `Generic` module, you must change the namespaces of specific models and mappers in the `Inventory` module. In the `Product.php` file, add the following line in the namespace section:

```
use Generic\Model\AbstractModel;
```

In the same way, in the `ProductTable` file, add the following line:

```
use Generic\Model\AbstractTable;
```

After these changes, you can execute **Behat run** again. As we only moved classes from one place to another, the result must be the same as that seen in the *Checking whether the product was inserted* section:

```
1 scenario (1 passed)
4 steps (4 passed)
0m0.35s (13.63Mb)
```

Note that the tests verify that what was working continues to work. Kent Beck, in his book *Test-Driven Development: By Example*, says that TDD eliminates the fear involved in development. Aided by tests, we can change our code and know immediately whether something is broken. With BDD, we keep this advantage of TDD and have files (`.feature`) that we can discuss directly with our customers.

You can make improvements securely with the aid of tests. For example, you must have noticed that the constructor is the same for the `Product` and `Discount` models and probably will be the same for the other models. Then, we can centralize this constructor in the `AbstractModel` class. Cut the method constructor from the `Product` class and paste it into `AbstractModel`. Then, delete the constructor of the `Discount` class. You can execute **Behat run** and check that everything continues to work.

Now, we have an appropriate generalization for our models and table mappers. As we can see, a Laminas module doesn't need to have a configuration or view folders. A module can have only what it needs.

Of course, we have implemented only one use case and we have created only two model classes, `Product` and `Discount`, but only `Product` is totally covered by tests. We need to create other models. But before we do, we have to prepare Behat to select scenarios. This is what we will do in the next section.

Creating other models with Behat

Let's create another feature file for the product updating use case. Inside the folder features, create the `productupdating.feature` file with this content:

---

productupdating.feature

```
Feature: Product updating
  In order to keep products up-to-date in the stock
  As an employee
  I need to be able to change data of a product from the
  product registration

  Scenario: Updating a product in the registration
    Given I have a product called "Troop Power Battery",
    which costs $2814
    When I change the name of product to "Enchanted Hammer"
    Then  there is a product in the table products called
    "Enchanted Hammer"
```

We need to have a class to be connected with this feature. We have created our first context class by copying. Now, let's use the **File** | **New** | **Class** menu to create the `ProductUpdatingContext` class. Select the `features/bootstrap` folder and the `Context` interface, as shown in *Figure 6.3*:

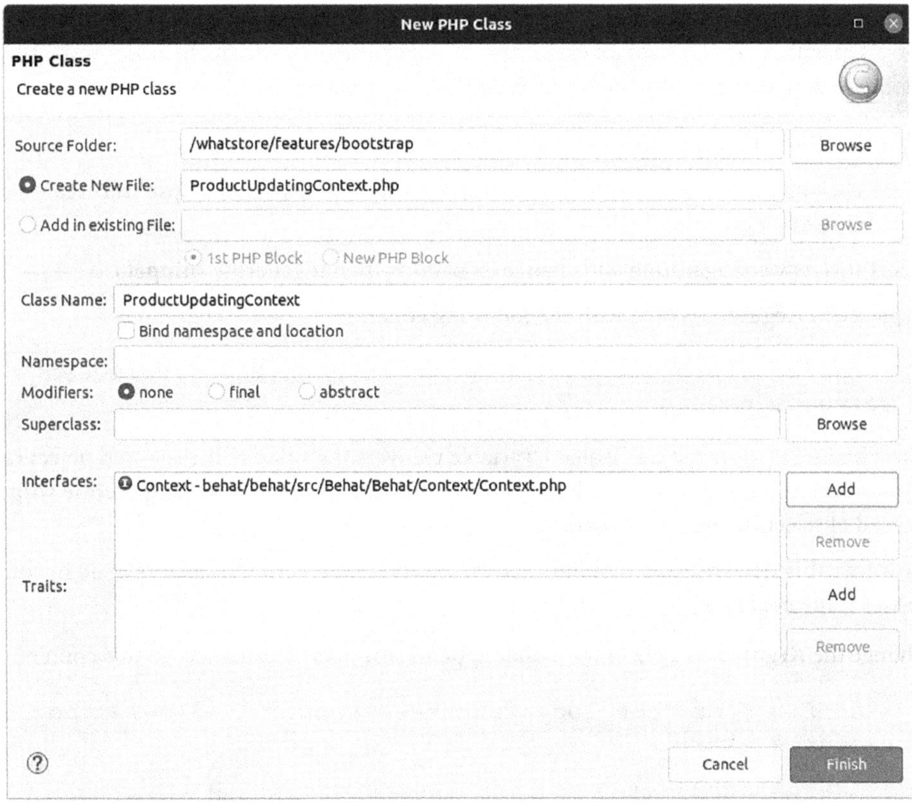

Figure 6.3 – Creating the ProductUpdatingContext class

We need to add the new context class to the `behat.yml` file in the `features` folder. The section contexts must be like this:

```
contexts:
    - ProductAdditionContext
    - ProductUpdatingContext
```

If you execute **Behat run** now, the output will end with the following:

```
[0] None
[1] ProductAdditionContext
[2] ProductUpdatingContext
```

Right, the new context class is appearing to be selected. But if you look at what appears before these options, you will notice that two scenarios are presented, *Product Addition* and *Product Updating*. When we don't tell Behat which scenarios we want, Behat brings all of them.

Right, but how can we tell Behat that we want to process only one scenario? It is easy. At this moment, we will do a separation. The **Behat run** external tool will continue to run the behavior tests. We will create another external tool configuration to show the code snippets:

1. Open the **Run | External Tools | External Tools Configurations** menu.

2. Right-click on the **Behat run** item, and select the **Duplicate** option. Eclipse will create a copy called `Behat run (1)`.

3. Select this new configuration and change its name to **Behat generate snippets**.

4. Replace the **Arguments** field with the following code:

    ```
    --config ${project_loc}/features/behat.yml  ${selected_
    resource_name}
    ```

    The `{select_resource_name}` variable recovers the value of the selected object (a file or a folder). From now on, we will select the feature file to execute **Behat generate snippets** instead of selecting the project name.

    However, this new external tool only shows the code. The item that inserts code in context classes is the next item.

5. Change the **Arguments** field of the **Behat append snippets** external tool to this content:

    ```
    --config ${project_loc}/features/behat.yml  ${selected_
    resource_name}
    ```

    Now, we are ready to see code generated for one specific scenario and to insert this code into the same scenario.

Select the `productupdating.feature` file and execute **Behat generate snippets** now. This time, only the feature described by `productupdating.feature` will be presented. Now, you can select option 2, which will refer to `ProductUpdatingContext`, and the **Console** tab will show the following snippets:

```
/**
 * @Given I have a product called :arg1, which costs
   $:arg2
 */
public function iHaveAProductCalledWhichCosts($arg1,
    $arg2)
{
    throw new PendingException();
}
```

```
    /**
     * @When I change the name of product to :arg1
     */
    public function iChangeTheNameOfProductTo($arg1)
    {
        throw new PendingException();
    }


    /**
     * @Then there is a product in the table products
       called :arg1
     */
    public function thereIsAProductInTheTableProductsCalled
       ($arg1)
    {
        throw new PendingException();
    }
```

As you can see, **Behat generate snippets** gives us the opportunity to check the generated code before adding it to the context class. Now, we can execute **Behat append snippets** to add the snippets.

> **Note about Behat definitions**
>
> Be careful when you write definitions in feature files. You shouldn't use equal definitions for different files. Behat compares definitions among files and warns you if there are equal definitions. For Behat, equal definitions in more than one feature file create an error, and the test will fail. It's because of this we have a "`there is a product called...`" definition in `ProductAddition.feature` and an "`I have a product called...`" definition in `productupdating.feature`. It is like Behat saying, "You already said this." Then, you answer, "Okay, I will use synonyms."

We can reuse some code from `ProductAdditionContext` in `ProductUpdatingContext`. Copy the `$product` and `$productTable` attributes and the constructor method from the first class to the second one.

As you have copied the constructor method, you may have noticed that this method depends on the private `getApplication` method. Do you also need to copy this method from `ProductAdditionContext`? No. But why?

The `getApplication` method will probably be necessary for several behavior test classes. If the same method appears in many different classes, this method can be centralized in a reusable unit of common code. In this case, this unit can be an abstract class.

We will create this class in our `Generic` module. So, create this class in the `Generic` module and a `Context` folder in the `src` folder. Then, create the abstract `AbstractContext` class in the `Generic/src/Context` folder with the **File** | **New** | **Class** menu. `AbstractContext` will implement the `Context` interface, common to behavior test classes. *Figure 6.4* shows how we must fill the fields of the class creation form:

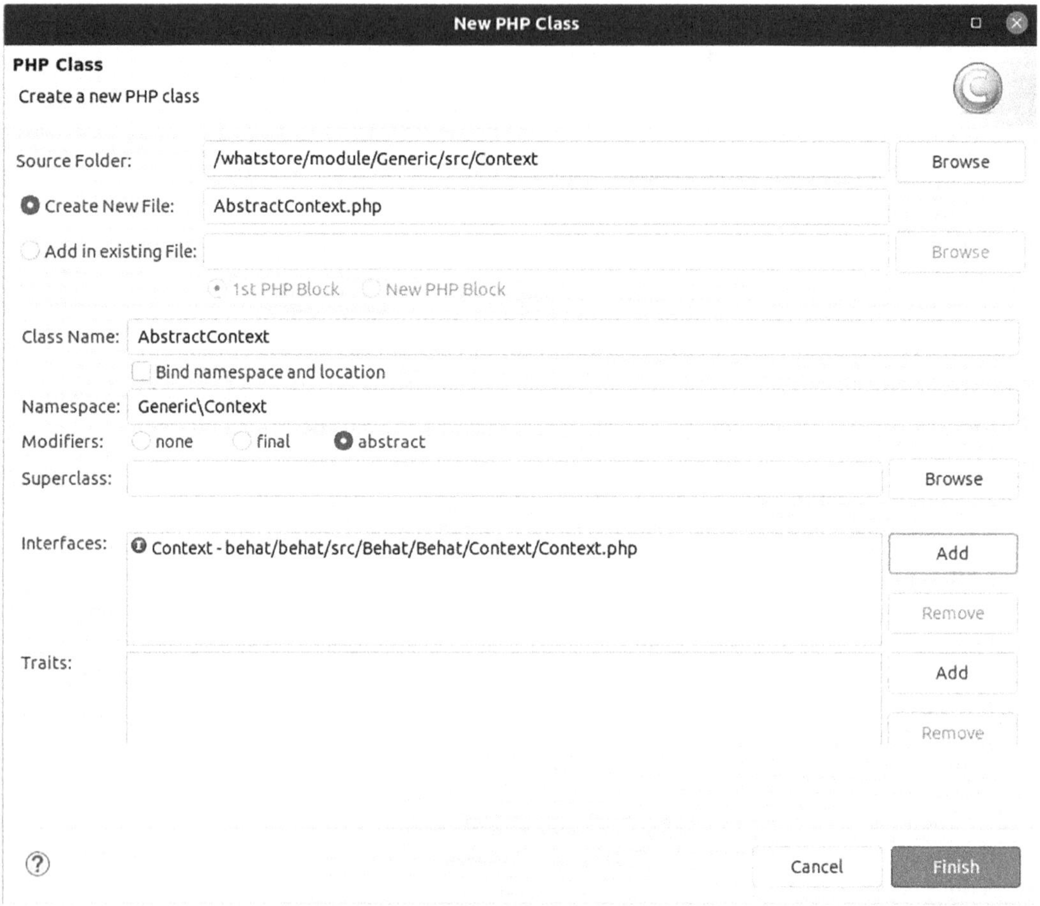

Figure 6.4 – Creating the abstract AbstractContext class

Since the `AbstractContext` class is created, the `getApplication` method from the `ProductAdditionContext` class is moved from `ProductAdditionContext` to the `AbstractContext` class. Change the scope of `getApplication` to `protected` because this method must be visible only for subclasses of `AbstractContext`. So, the `AbstractContext` class will be like this:

AbstractContext.php

```php
<?php
namespace Generic\Context;

use Behat\Behat\Context\Context;
use Laminas\Stdlib\ArrayUtils;
use Laminas\Mvc\Application;

abstract class AbstractContext implements Context
{
    protected function getApplication()
    {
        $appConfig = require __DIR__ . '/../../../../config
            /application.config.php';
        if (file_exists(__DIR__ . '/../../../../config/
            development.config.php')) {
            $appConfig = ArrayUtils::merge($appConfig,
                require __DIR__ . '/../../../../config/
                    development.config.php');
        }

        return Application::init($appConfig);
    }
}
```

Of course, we had to change the lines that have file paths because the relative level is different. Before, we had three levels to return to root path; now, we have four levels.

From now on, the `ProductAdditionContext` and `ProductUpdatingContext` classes must extend `AbstractContext` instead of implementing the `Context` interface. Make this change on the declaration of these classes, according to this example for `ProductAdditionContext`:

```php
class ProductAdditionContext extends AbstractContext
```

After the adjustments related to inheritance, we can conclude the implementation in the `ProductUpdatingContext` class. Let's show the code of this class step by step. First, as we always have to do in PHP classes, we need to add the references for the dependencies with the `use` command:

```php
<?php
use Behat\Behat\Tester\Exception\PendingException;
use Behat\Gherkin\Node\PyStringNode;
use Behat\Gherkin\Node\TableNode;

use Inventory\Model\Product;
use Inventory\Model\ProductTable;
use Inventory\Model\ProductTableFactory;
use PHPUnit\Framework\Assert;
use Generic\Context\AbstractContext;
```

Note that Behat has classes that are an implementation of the Gherkin language. These classes, specifically, are related to the parsing of the `.feature` files. In sequence, we declare the `ProductUpdating` class and two private attributes:

```php
/**
 * Defines application features from the specific context.
 */
class ProductUpdatingContext extends AbstractContext
{
    private ?Product $product = null;
    private ?ProductTable $productTable = null;
```

We have defined an attribute for the model object (`$product`), which keeps the data in memory, and another attribute for the table gateway object (`$productTable`), which recovers data from a table to memory and saves data from memory to a table. In sequence, we defined the constructor method, executed automatically when an instance of the `ProductUpdating` class is created:

```php
/**
 * Initializes context.
 *
 * Every scenario gets its own context instance.
 * You can also pass arbitrary arguments to the
 * context constructor through behat.yml.
```

```
    */
    public function __construct()
    {
        $this->product = new Product();
        $this->productTable = $this->getApplication()->
            getServiceManager()->get('ProductTable');
    }
```

The `ProductUpdating` constructor instantiates the `Product` class directly and uses a factory to instantiate the `ProductTable` class. In sequence, we have a method that implements the Gherkin phrase annotated with `@Given`:

```
    /**
     * @Given I have a product called :arg1, which costs
         $:arg2
     */
    public function iHaveAProductCalledWhichCosts($arg1,
        $arg2)
    {
        $data = [
            'name'  => $arg1,
            'price' => $arg2
        ];
        $this->product->exchangeArray($data);
    }
```

The `iHaveAProductCalledWhichCosts` method populates the instance of the `Product` class with data provided by the `productupdating.feature` file. In sequence, we have a method that implements the Gherkin phrase annotated with `@When`:

```
    /**
     * @When I change the name of product to :arg1
     */
    public function iChangeTheNameOfProductTo($arg1)
    {
        $product = $this->productTable->getByField
            ('name',$this->product->name);
        $product->name = $arg1;
```

```
    $this->productTable->save($product);
}
```

The iChangeTheNameOfProductTo method recovers a product by its name; change this name and update the record in the table. In sequence, we have a method that implements the Gherkin phrase annotated with @Then:

```
/**
 * @Then there is a product in the table products
   called :arg1
 */
public function thereIsAProductInTheTableProductsCalled
    ($arg1)
{
    $this->product = $this->productTable->getByField
        ('name',$arg1);

    Assert::assertEquals($arg1, $this->product->name);
}
```

The thereIsAProductInTheTableProductsCalled method represents the connection between BDD and TDD. In this method, we use PHPUnit to make an assertion. Finally, we revert the changes executed by previous methods with the destructor method:

```
public function __destruct()
{
    $product = $this->productTable->getByField(
        'name',$this->product->name);
    $this->productTable->delete($product->code);
    $product = $this->productTable->getByField
        ('name',$product->name);
    Assert::assertEmpty($product->name);
}
```
}

You can see that most of the implementation of ProductUpdatingContext is similar to what we have done for the ProductAdditionContext class. Observe that we have added an assertion in the destructor method to check whether the product was removed. So, there is no necessity to create a feature file for the use case product deletion. In fact, you can organize the features for grouping use cases and, therefore, avoid repetitions.

With this, we have concluded what you need to know about model creation for Laminas projects aided by Behat.

Summary

In this chapter, we have provided an introduction to BDD with Behat. We have learned that BDD allows us to build on the useful skills of TDD and also gives us a way to write user stories that generate part of our test code.

With the aid of Behat as a BDD tool for PHP, we learned how to create models and table mappers from user stories. In fact, we saw that the combination of Behat and Laminas is very powerful for the construction of the model layer.

We also learned to create a `Generic` module with abstract classes to serve other modules and reuse generic classes created in *Chapter 4, From Object-Relational Mapping to MVC Containers*. The same `Generic` module was used to store an abstraction for context classes, the implementations of the behavior test. Here, we learned that it is useful to have abstractions to reuse, avoiding the unnecessary work and complexity generated by replication. As this act of abstraction occurs for every code unit, abstract classes can grow to facilitate more code reuse during each evolution stage of a software project. Then, each new feature applied to more than one class is a warning to reduce code lines using abstractions.

In the next chapter, we will create the control and view layers of our application, which will use the models and table mappers created in this chapter.

7

Request Control and Data View

This chapter will demonstrate how to use the Laminas implementations for the controller and view layers in an MVC application. We will start with a brief review of the relationship between PHP and **Hypertext Transfer Protocol** (**HTTP**). We will look at how Laminas deals with HTTP requests and responses. After that, we will implement the controllers and views of the `Inventory` module, that is, we will learn how a module works in the Laminas MVC. Finally, we will learn how to create a page controller and page templates, along with learning about how a page controller works.

In this chapter, we'll be covering the following topics:

- Understanding the relationship between HTTP and PHP
- The request life cycle in the Laminas MVC
- Implementing CRUD with a controller and view pages

By the end of this chapter, you will be able to implement a registration using a Laminas module with an MVC pattern and you will understand how the `laminas-mvc` component handles HTTP requests. In addition, you will know how the Laminas controller page classes work and you will be able to manage HTML pages with Laminas modules.

Technical requirements

All the code related to this chapter can be found at `https://github.com/PacktPublishing/PHP-Web-Development-with-Laminas/tree/main/chapter07/whatstore`.

Understanding the relationship between HTTP and PHP

The essence of HTTP is this: a computer called the client sends a text message (or a request) to another computer called the server. The server replies with a text message (or a response). HTTP has been defined by RFCs 7230 to 7237 since 2014. According to RFC 7230, "client" and "server" are roles, so the same computer can be a client and a server. In a distributed system, based on microservices, for instance, each microservice can act as a client for a microservice as well as a server for another one.

Some programming languages were adapted to build web applications. PHP, however, is a language that was originally created to work with HTTP. Rasmus Lerdorf developed PHP because he was not satisfied with the tools available for creating dynamic pages, or more precisely, HTML pages integrated with databases. PHP has evolved in parallel to the protocols used for web application development. It is very fast to make dynamic pages available with PHP. The challenge is not to complicate a PHP application and keep the application under control as it grows.

Laminas has a component that encapsulates HTTP elements as objects. This is the `laminas-http` component. This component is used by `laminas-mvc`. The `laminas-http` component implements interfaces defined by `laminas-stdlib`. The `laminas-stdlib` component represents the highest level of reuse for the Laminas framework, with interfaces and abstract classes used by most of the Laminas components.

We will see how `laminas-mvc` uses classes from `laminas-http`. In the following section, we will learn about the request life cycle in a Laminas MVC application.

The request life cycle in the Laminas MVC

At this moment, you can quickly launch the *whatstore* application and see the home page with the embedded web server for PHP. From the `whatstore` directory, you can type the following command:

```
php -S localhost:8000 -t public/
```

When you open `localhost:8000` in your browser, you will see the same welcome page that you will have seen in *Chapter 2, Setting Up the Environment for Our E-Commerce Application*. Of course, we haven't changed any view yet.

In *Chapter 5, Creating the Virtual Store Project*, we learned a little about the `public` directory. We learned that the structure of an MVC Laminas application is like that of a medieval castle, where the only way to enter is over the drawbridge. The drawbridge here is the `public` directory. All HTTP requests are passed to the `index.php` file inside the `public` directory. So, what does `index.php` do? It does this:

- Imports the `autoload.php` Composer file.

- Checks whether the Laminas `Application` class exists.

- Retrieves the configuration. If there is a development configuration, this configuration overrides the production configuration.

- Instantiates the `Application` class using the `Application::init` method. This instance receives objects from `laminas-http` via dependency injection.

- Executes the `Application->run` method from the created instance.

When `index.php` executes `Application->run`, an **inversion of control** happens. From this point onward, Laminas assumes control and will call your specific code, or, in other words, your business code.

> Note: inversion of control
>
> You can read more about the inversion of control pattern in the following article by Martin Fowler: `https://martinfowler.com/bliki/InversionOfControl.html`.

Figure 7.1 gives a general view of the life cycle of a request in a Laminas MVC application:

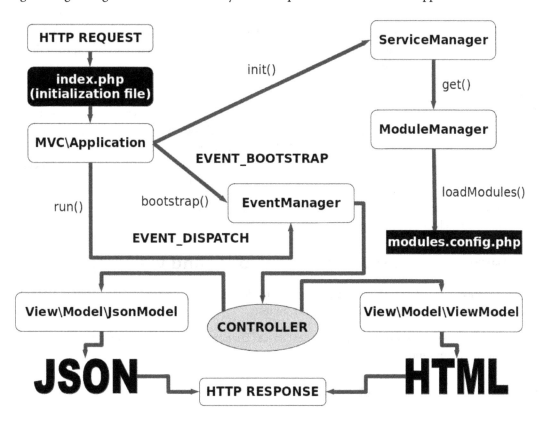

Figure 7.1 – HTTP request life cycle in a Laminas MVC application

Laminas uses event-driven programming for processing the steps of the request life cycle: bootstrapping, routing, dispatching the controller, and rendering the view. We can divide this life cycle into two main moments:

- When the `Application::init` method is executed, `ServiceManager` gets an instance of `ModuleManager`, which loads the modules registered in the `modules.config.php` file. The `Application::init` method also triggers the `EVENT_BOOTSTRAP` event using the `bootstrap` method. The `EventManager` class notifies the listeners of this event for executing the bootstrap tasks.

- When the `Application->run` method is executed, `Application` triggers the `EVENT_DISPATCH` event. The `EventManager` class notifies the listeners of this event and one of them instantiates a controller class. This controller class, in turn, renders a view for the HTTP response. This view can be a JSON object, an HTML page, or potentially any other format that can be consumed by another application.

Wait a moment. Which controller will be called, exactly? It depends on the route. As we already said in *Chapter 4*, *From Object-Relational Mapping to MVC Containers*, Laminas defines routes in its modules. Routes are associations between URL patterns and controller classes. The `Application->run` method also triggers an event called `EVENT_ROUTE`, and `EventManager` notifies certain listeners related to `ModuleManager` (`ModuleRouteListener` and `RouteListener`). These listeners define the controller to be called from the route that matches the URL.

We will understand this life cycle better in the next section, where we will implement a controller and view pages for product registration.

Implementing CRUD with a controller and view pages

We will create a controller in the `Inventory` module for managing products. Following the good practice of TDD, we will create this controller from tests.

In this section, we will first test the current test cases. Then, we will create the tests for inserting, recovering, updating, and deleting products.

Testing our current test cases

The first step is to check the current state of our test cases by following these steps:

1. Select the `whatstore` project and right-click on **Run As/PHPUnit Test**. You will see the output of PHPUnit in the **PHPUnit** tab, as *Figure 7.2* shows:

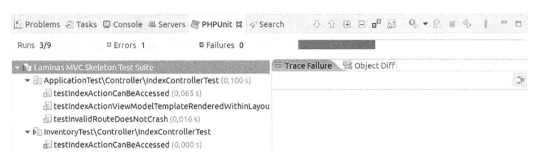

Figure 7.2 – PHPUnit output

As you can see, the `IndexControllerTest` class from the `Inventory` module has failed. It failed because this class is a copy of `IndexControllerTest` from the `Application` module and the routes and pages are different. Let us fix this class.

2. First, change the `testIndexActionCanBeAccessed` method for `Inventory\Controller\IndexControllerTest` to the following code:

```php
public function testIndexActionCanBeAccessed(): void
{
    $this->dispatch('/inventory', 'GET');
    $this->assertResponseStatusCode(200);
    $this->assertModuleName('inventory');
    $this->assertControllerName(IndexController::
        class); // as specified in router's
            controller name alias
    $this->assertControllerClass
        ('IndexController');
    $this->assertMatchedRouteName('inventory');
}
```

3. Then, remove the `testIndexActionViewModelTemplateRenderedWithinLayout` method from `InventoryTest\Controller\IndexControllerTest`. This method only works for *whatstore* home page.

4. Remove the `testInvalidRouteDoesNotCrash` method too. This method is already defined for the `Application` module.

    If you execute **Run As/PHPUnit Test** again, `IndexControllerTest` fails again. Now, the reason is that the `inventory` route does not exist yet. We need to create this route in the `module.config.php` file of the `Inventory` module.

5. Add the following code to the `module.config.php` file so that the `routes` key in the file looks as follows:

```
'routes' => [
    'inventory' => [
        'type'    => Segment::class,
        'options' => [
            'route'    => '/inventory
            [/:controller[/:action[/:key]]]',
            'defaults' => [
                'controller' => Controller
                \IndexController::class,
                'action' => 'index'
            ],
        ],
    ],
],
```

We have created a route of the `Segment` kind, a route that accepts arguments. We already learned about this route in *Chapter 4, From Object-Relational Mapping to MVC Containers*. The words between brackets are optional. The words preceded by a colon are variables. The `inventory` route allows us to define the controller to be instantiated, the action to be executed, and a `key` argument for selecting the object to be handled. In case these arguments are not informed, the default controller is `IndexController` and the default `action` is `index`.

Let us return to tests. If you execute **Run As/PHPUnit Test** one more time, the `IndexControllerTest` class from the `Inventory` module will run successfully. However, the `IndexControllerTest` class from the `Store` module will fail. No problem.

6. Change the `testIndexActionCanBeAccessed` method for the `StoreTest\Controller\IndexControllerTest` class to the following code:

```php
public function testIndexActionCanBeAccessed():
void
{
    $this->dispatch('/store', 'GET');
    $this->assertResponseStatusCode(200);
    $this->assertModuleName('store');
    $this->assertControllerName
        (IndexController::class); // as specified
            in router's controller name alias
```

```
            $this->assertControllerClass
                ('IndexController');
            $this->assertMatchedRouteName('store');
        }
```

7. Then, according to what we already have done, remove the testIndexActionViewMod-elTemplateRenderedWithinLayout method from StoreTest\Controller\IndexControllerTest. Don't forget to remove the testInvalidRouteDoesNot-Crash method too.

As we should have learned, we need to create this route in the module.config.php file of the Store module. The routes key in this file must be as follows:

```
'routes' => [
    'store' => [
        'type'    => Segment::class,
        'options' => [
            'route'   => '/store
            [/:controller[/:action[/:key]]]',
            'defaults' => [
                'controller' => Controller
                \IndexController::class,
                'action'     => 'index',
            ],
        ],
    ],
],
```

Great – for a while, our tests are running successfully. From now on, we will implement the controller and view layers for products using tests. Our controller layer for products will have two controller classes, one for receiving requests from an HTML default page, part of the whatstore application, and another class for receiving requests from any client. This architecture allows other clients, such as mobile applications, to consume services from whatstore. *Figure 7.3* depicts that service-oriented architecture:

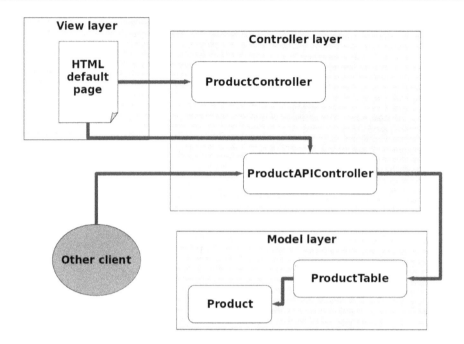

Figure 7.3 – Architecture for product controllers

We will start with `ProductAPIController`, which can be consumed by any client. As we have defined, we will implement a test class for constructing the controller class. Before we do that, do the following:

1. In the `test/Controller` folder of the `Inventory` module, copy the `IndexController-Test.php` file to `ProductAPIControllerTest.php`.

2. Change the class name in this last file to `ProductAPIControllerTest` and remove the `testIndexActionCanBeAccessed` method.

Now, let's implement the test methods for the `ProductAPIController` class.

Testing product insertion

The first test method to create in `ProductAPIControllerTest` will be `testProductionInsert`. The code will be as follows:

```php
public function testProductInsert(): void
{
    $discount = new Discount();
    $data = [
```

```
            'name' => 'no discount',
            'operator' => '-',
            'factor' => 0
        ];
        $discount->exchangeArray($data);
        $discountTable = $this->getApplication()->
            getServiceManager()->get('DiscountTable');
        $discountTable->save($discount);
        $discount = $discountTable->getByField
            ('name','no discount');
        $_POST = [
            'name' => 'Cosmic Cube',
            'price' => 5642,
            'code_discount' => $discount->code
        ];
        $this->dispatch('/inventoryapi/productapi', 'POST',
            $_POST);
        $this->assertResponseStatusCode(200);
        $this->assertModuleName('inventory');
        $this->assertControllerName('productapi'); // as
            specified in router's controller name alias
        $this->assertControllerClass
            ('ProductAPIController');
        $this->assertMatchedRouteName('inventoryapi');
    }
```

The test ProductInsert method does the following:

- Creates and persists a Discount object
- Dispatches an HTTP POST request for the /inventoryapi/productapi endpoint
- Checks whether the response has the 200 status (success)
- Checks whether the called module was Inventory
- Checks whether the alias for the called controller name was productapi
- Checks whether the called controller name was ProductAPIController
- Checks whether the route name was inventoryapi

We need to define the `inventoryapi` route, which redirects to `ProductAPIController`. We will also need to define another route for `ProductController`. So, in the `modules.config. php` file of the `Inventory` module, we will add the following routes to the `routes` key:

```
'routes' => [
    'inventory' => [
        'type'     => Segment::class,
        'options' => [
            'route'     => '/inventory
                [/:controller[/:action[/:key]]]',
            'defaults' => [
                'controller' => Controller
                    \IndexController::class,
                'action' => 'index'
            ],
        ],
    ],
    'inventoryapi' => [
        'type'     => Segment::class,
        'options' => [
            'route'     => '/inventoryapi
                [/:controller[/:key]]'
        ],
    ],
],
```

> **Important note: RESTful controllers must not have a default action**
>
> We must not define a default action when we use a RESTful controller. The reason is that the default action overrides the inference of the HTTP method.

After implementing the test for product insertion, we will implement the test for product recovery next.

Testing product recovery

We will implement the testProductionRecover method in the ProductAPIControllerTest class and the code should look as follows:

```
public function testProductRecover(): void
{
    $name = 'Cosmic Cube';
    $this->dispatch('/inventoryapi/productapi/' .
        $name, 'GET');
    $this->assertResponseStatusCode(200);
    $this->assertModuleName('inventory');
    $this->assertControllerName('productapi'); // as
        specified in router's controller name alias
    $this->assertControllerClass
        ('ProductAPIController');
    $this->assertMatchedRouteName('inventoryapi');
    $body = $this->getResponse()->getBody();
    $this->assertStringContainsString($name, $body);
}
```

The testProductRecover method does the following:

- Dispatches an HTTP GET request for the /inventoryapi/productapi endpoint (observe that the argument to search is the product name)

- Checks whether the response has the 200 status (success)

- Checks whether the called module was Inventory

- Checks whether the alias for called controller name was productapi

- Checks whether the called controller name was ProductAPIController

- Checks whether the route name was inventoryapi

- Checks whether the response body has the product name

After implementing the test for product recovery, we will implement the test for product updates.

Testing product updates

Now, we will implement the testProductUpdate method in the class ProductAPIControllerlerTest. The code will look as follows:

```
public function testProductUpdate(): void
{
    $productTable = $this->getApplication()->
        getServiceManager()->get('ProductTable');
    $product = $productTable->getByField('name',
        'Cosmic Cube');
    $product->name = 'Infinity Gauntlet';
    $product->price = 9654;
    $_POST = $product->toArray();
    $this->dispatch('/inventoryapi/productapi/' .
        $product->code, 'PUT', $_POST);
    $this->assertResponseStatusCode(200);
    $this->assertModuleName('inventory');
    $this->assertControllerName('productapi'); // as
        specified in router's controller name alias
    $this->assertControllerClass
        ('ProductAPIController');
    $this->assertMatchedRouteName('inventoryapi');
}
```

The testProductUpdate method does the following:

- Gets an instance of ProductTable
- Recovers the product inserted before
- Extracts the product data to the $_POST variable
- Dispatches an HTTP PUT request for the /inventoryapi/productapi endpoint (observe that the argument to search is the product code)
- Checks whether the response has the 200 status (success)
- Checks whether the called module was Inventory
- Checks whether the alias for called controller name was productapi

- Checks whether the called controller name was `ProductAPIController`

- Checks whether the route name was `inventoryapi`

After implementing the test for the product updates, we will implement the test for product deletion.

Testing product deletion

Finally, to finish the API test, we will implement the `testProductionDelete` method in the `ProductAPIControllerTest` class. The code should look as follows:

```
public function testProductDelete(): void
{
    $productTable = $this->getApplication()->
        getServiceManager()->get('ProductTable');
    $product = $productTable->getByField('name',
        'Infinity Gauntlet');
    $this->dispatch('/inventoryapi/productapi/' .
        $product->code, 'DELETE');
    $this->assertResponseStatusCode(200);
    $this->assertModuleName('inventory');
    $this->assertControllerName('productapi'); // as
        specified in router's controller name alias
    $this->assertControllerClass
        ('ProductAPIController');
    $this->assertMatchedRouteName('inventoryapi');
    $discountTable = $this->getApplication()->
        getServiceManager()->get('DiscountTable');
    $discount = $discountTable->getByField('name','no
        discount');
    $discountTable->delete($discount->code);
}
```

The `testProductDelete` method does the following:

- Gets an instance of `ProductTable`

- Recovers the product inserted before

- Dispatches an HTTP DELETE request for the `/inventoryapi/productapi` endpoint (observe that argument to delete is the product code)

- Checks whether the response has the 200 status (success)
- Checks whether the called module was Inventory
- Checks whether the alias for called controller name was productapi
- Checks whether the called controller name was ProductAPIController
- Checks whether the route name was inventoryapi
- Recovers the Discount object inserted by the testProductInsert method
- Uses a DiscountTable instance to delete the discount

If you execute **Run As/PHPUnit Test** now, ProductAPIControllerTest will fail, because we don't have the ProductAPIController class. We will create this class in the following section.

Creating the ProductAPIController class

Create a ProductAPIController.php file in the src/Controller folder of the Inventory module. Let us see, part by part, the source code for the ProductAPIController.php file. This file begins with the following content:

```php
<?php
namespace Inventory\Controller;
use Laminas\Mvc\Controller\AbstractRestfulController;
use Laminas\View\Model\JsonModel;
use Inventory\Model\Product;
use Inventory\Model\ProductTable;
```

In the first part of the ProductAPIController class, we use four dependencies. The ProductAPIController class extends AbstractRestfulController. This superclass is designed to map the HTTP methods to specific methods:

```php
class ProductAPIController extends
AbstractRestfulController
{
```

The $identifierName attribute, inherited from AbstractRestfulController, is overridden to have the key value. The default value for $identifierName is *id*, but the inventoryapi route defines the *key* argument.

As ProductAPIController handles products, it needs to have an attribute for a ProductTable instance:

```php
    protected $identifierName = 'key';
    private ProductTable $productTable;
```

The constructor method receives an instance of `ProductTable` and assigns it to the attribute `productTable`:

```
public function __construct(ProductTable $productTable)
    {
        $this->productTable = $productTable;
    }
```

HTTP POST is mapped to the `create` method:

```
public function create($data)
    {
        $product = new Product();
        $product->exchangeArray($data);
        $inserted = $this->productTable->save($product);
        return new JsonModel(['inserted' => $inserted]);
    }
```

The `$data` argument, received by the `create` method, contains the data from `$_POST`.

HTTP GET (with an argument) is mapped to the `get` method:

```
    public function get($id)
    {
        $field = (is_numeric($id) ? 'code' : 'name');
        $id = (is_numeric($id) ? (int) $id : $id);
        $product = $this->productTable->getByField($field,
            $id);
        return new JsonModel(['product' => $product->
            toArray()]);
    }
```

HTTP PUT is mapped to the `update` method:

```
public function update($id, $data)
    {
        $product = $this->productTable->getByField
            ('code', $id);
        $product->exchangeArray($data);
        $updated = $this->productTable->save($product);
```

```
    return new JsonModel(['updated' => $updated]);
}
```

HTTP DELETE is mapped to the delete method:

```
public function delete($id)
{
    $deleted = $this->productTable->delete($id);
    return new JsonModel(['deleted' => $deleted]);
}
}
```

Observe that the methods from ProductAPIController return an instance of JsonModel.

The JsonModel class converts the data injected into its constructor for JSON format – an appropriate behavior for an API. However, to use JsonModel to send data to the view layer, we need to set a strategy. The default strategy is for HTML page rendering. To use the JSON format in responses, we must add the following item to the view_manager key in modules.config.php in the Inventory module:

```
'strategies' => [
    'ViewJsonStrategy',
],
```

For the ProductAPIController class to be invoked, we need to change the key controllers in modules.config.php in Inventory to the following content:

```
'controllers' => [
    'aliases' => [
        'productapi' => Controller\
            ProductAPIController::class
    ],
    'factories' => [
        Controller\IndexController::class =>
            InvokableFactory::class,
        Controller\ProductAPIController::class =>
          Controller\ProductAPIControllerFactory::class
    ],
],
```

The aliases key, according to what we have seen in *Chapter 4, From Object-Relational Mapping to MVC Containers*, is used to map controller aliases to real controller class names. The factories key, as we know, defines how the controller classes will be created. We need to create the ProductAPIControllerFactory class in the src/Controller folder of the Inventory module. The source code will look as follows:

ProductAPIControllerFactory.php

```php
<?php
namespace Inventory\Controller;
use Laminas\ServiceManager\Factory\FactoryInterface;
use Interop\Container\ContainerInterface;
class ProductAPIControllerFactory implements
FactoryInterface
{
    public function __invoke(ContainerInterface $container,
        $requestedName, array $options = null)
    {

        $productTable = $container->get('ProductTable');
        return new ProductAPIController($productTable);
    }
}
```

If you execute **Run As/PHPUnit Test** now, the tests will run successfully. So, the API for the products seems complete. In the following section, we will create the class responsible for controlling the HTML pages in our whatstore application.

Implementing the ProductController class

We will now create a ProductControllerTest class in the Inventory module. Let us do this, part by part, starting with the following source code:

```php
<?php
declare(strict_types=1);
namespace InventoryTest\Controller;
use Inventory\Controller\IndexController;
use Laminas\Stdlib\ArrayUtils;
use Laminas\Test\PHPUnit\Controller
\AbstractHttpControllerTestCase;
use Inventory\Controller\ProductController;
```

In the first part of the `ProductAPIController` class, we used four dependencies.

Next, we will use the `setUp` method to prepare the test environment, recovering the application configuration:

```
class ProductControllerTest extends
    AbstractHttpControllerTestCase
{
    public function setUp(): void
    {
        $configOverrides = [];
        $this->setApplicationConfig(ArrayUtils::merge(
            include __DIR__ . '/..//..//../config/
                application.config.php',
            $configOverrides
        ));
        parent::setUp();
    }
}
```

Observe that the content of the application configuration is merged with an `$configOverrides` array. This `$configOverrides` array allows us to change the configuration values for running tests.

Next, we will use the `testIndexActionCanBeAccessed` method to test the request for the index page of `ProductController`, which will list the products:

```
public function testIndexActionCanBeAccessed(): void
{
    $this->dispatch('/inventory/product', 'GET');
    $this->assertResponseStatusCode(200);
    $this->assertModuleName('inventory');
    $this->assertControllerName('product'); // as
        specified in router's controller name alias
    $this->assertControllerClass('ProductController');
    $this->assertMatchedRouteName('inventory');
}
```

Lastly, we will use the `testEditActionCanBeAccessed` method to test the request for the edit page, which needs to allow us to insert and update products:

```php
public function testEditActionCanBeAccessed(): void
{
    $this->dispatch('/inventory/product/edit', 'GET');
    $this->assertResponseStatusCode(200);
    $this->assertModuleName('inventory');
    $this->assertControllerName('product'); // as
        specified in router's controller name alias
    $this->assertControllerClass('ProductController');
    $this->assertMatchedRouteName('inventory');
}
}
```

Now, let us create the class to be tested. The following code block shows the class that controls the HTML interface for product registration:

ProductController.php

```php
<?php
declare(strict_types=1);
namespace Inventory\Controller;
use Laminas\Mvc\Controller\AbstractActionController;
use Laminas\View\Model\ViewModel;
use Inventory\Model\ProductTable;
use Inventory\Model\DiscountTable;
class ProductController extends AbstractActionController
{
    private ?ProductTable $productTable = null;
    private ?DiscountTable $discountTable = null;
    public function __construct(ProductTable $productTable,
        DiscountTable $discountTable)
    {
        $this->productTable = $productTable;
        $this->discountTable = $discountTable;
    }
    public function indexAction()
```

```
    {
        $products = $this->productTable->getAll();
        return new ViewModel(['products' => $products]);
    }
    public function editAction()
    {
        $key = $this->params('key');
        $product = $this->productTable->
            getByField('code',$key);
        $discounts = $this->discountTable->getAll();
        return new ViewModel([
            'product' => $product,
            'discounts' => $discounts
        ]);
    }
}
```

> **An array in place of ViewModel**
>
> You also can return an array in place of a `ViewModel` object. The controller class will inject the array into a `ViewModel` object.

The `indexAction` method recovers a set of products and delivers it to the view layer. The `editAction` method tries to find a product by its code and delivers the product object and a set of discounts to the view layer. As you may be able to tell, the recovery of products and discounts depends on the table mapper objects. These objects are injected by the constructor method. This injection is made by a factory class, `ProductControllerFactory`, whose source code looks as follows:

ProductControllerFactory.php

```php
<?php
namespace Inventory\Controller;
use Laminas\ServiceManager\Factory\FactoryInterface;
use Interop\Container\ContainerInterface;
class ProductControllerFactory implements FactoryInterface
{
    public function __invoke(ContainerInterface $container,
        $requestedName, array $options = null)
```

```
    {
        $productTable = $container->get('ProductTable');
        $discountTable = $container->get('DiscountTable');
        return new ProductController($productTable,
            $discountTable);
    }
}
```

We need to register this factory in modules.config.php. In addition, we also need to register an alias for ProductController. The controllers key in modules.config.php in the Inventory module will be as follows:

```
'controllers' => [
    'aliases' => [
        'productapi' => Controller\
            ProductAPIController::class,
        'product' => Controller\
            ProductController::class
    ],
    'factories' => [
        Controller\IndexController::class =>
            InvokableFactory::class,
        Controller\ProductAPIController::class =>
        Controller\ProductAPIControllerFactory::class,
        Controller\ProductController::class =>
            Controller\ProductControllerFactory::class,
    ],
],
```

If you execute **Run As/PHPUnit Test** now, the tests will fail, presenting the following messages:

```
There were 2 failures:
1) InventoryTest\Controller\
ProductControllerTest::testIndexActionCanBeAccessed
Failed asserting response code "200", actual status code is
"500"
Exceptions raised:
Exception 'Laminas\View\Exception\RuntimeException' with
message 'Laminas\View\Renderer\PhpRenderer::render: Unable
```

```
to render template "inventory/product/index"; resolver could
not resolve to a file' in /opt/lampp/htdocs/whatstore/vendor/
laminas/laminas-view/src/Renderer/PhpRenderer.php:511

/opt/lampp/htdocs/whatstore/vendor/laminas/laminas-test/src/
PHPUnit/Controller/AbstractControllerTestCase.php:434

/opt/lampp/htdocs/whatstore/module/Inventory/test/Controller/
ProductControllerTest.php:33

2) InventoryTest\Controller\
ProductControllerTest::testEditActionCanBeAccessed

Failed asserting response code "200", actual status code is
"500"

Exceptions raised:

Exception 'Laminas\View\Exception\RuntimeException' with
message 'Laminas\View\Renderer\PhpRenderer::render: Unable
to render template "inventory/product/edit"; resolver could
not resolve to a file' in /opt/lampp/htdocs/whatstore/vendor/
laminas/laminas-view/src/Renderer/PhpRenderer.php:511

/opt/lampp/htdocs/whatstore/vendor/laminas/laminas-test/src/
PHPUnit/Controller/AbstractControllerTestCase.php:434

/opt/lampp/htdocs/whatstore/module/Inventory/test/Controller/
ProductControllerTest.php:43
```

Observe that the error messages talk about template rendering. The errors occur because we haven't created HTML pages for generating the HTTP response for these actions. Let us create them in the next section.

Implementing the pages for ProductController

By default, when a controller method returns a `ViewModel` object, the view manager will search for a page with the same name as the method (except for the `Action` suffix). If the previous tests have failed for the `indexAction` and `editAction` methods, then we must create `index.phtml` and `edit.phtml` files. These files must be created in the `Inventory/view/inventory/product` folder, as we can see in *Figure 7.4*. `.phtml` files are PHP files:

Figure 7.4 – View pages for ProductController in the Inventory module

The simple existence of the edit.phtml and index.phtml files is enough for ProductControllerTest to run successfully, but this does not mean that a user can interact with the web pages. The ProductControllerTests tests are for checking whether these files exist (and as a consequence, your content is merged with the layout to compose the response).

First, add the following code to the index.phtml file:

index.phtml

```
<h1>Product Registration</h1>
<a href="<?=$this->url('inventory', ['controller' =>
    'product','action' => 'edit'])?>">Add product</a>
<table>
    <thead>
    <tr>
            <th>code</th>
            <th>name</th>
            <th>product</th>
            <th>discount</th>
```

```
        </tr>
        </thead>
        <tbody>
<?php
foreach ($this->products as $product) :
?>
        <tr>
            <td><button id="<?=$product->code?>"><?=$product-
>code?>
</button></td>
    <td><?=$product->name?></td>
    <td><?=$product->price?></td>
                <td><?=$product->discount->name?>
</td>
                <td><button id="<?=$product->discount-
>name?>">delete</button></td>
            </tr>
<?php
endforeach;
?>
        </tbody>
</table>
```

The $this->products attribute is the same object passed in the constructor of the ViewModel instance in the ProductController->indexAction method. The $this keyword in the .phtml files refers to the Laminas\View\Renderer\PhpRenderer class. This class has helpers to generate view elements and contains the data sent by the controller. The index.phtml file will render the page, as seen in *Figure 7.5*:

Product Registration

Add product

code name price discount

© 2022 Laminas Project a Series of LF Projects, LLC.

Figure 7.5 – Product list page

Add product takes you to the page rendered by `edit.phtml`. The content of this file will be the following:

edit.phtml

```
<h1>Product <?=(empty($this->product->code) ? 'insert' :
    'update')?></h1>
<form id="product">
    Name: <input type="text" name="name" autofocus=
        "autofocus" value="<?=$this->product->name?>"><br/>
    Price: <input type="text" name="price" value="<?=$
        this->product->price?>"><br/>
    Discount: <select name="code_discount" value="<?=$
        this->product->discount->code?>">
<?php
foreach($this->discounts as $discount):
?>
<option value="<?=$discount->code?>" <?=($discount->code ==
    $this->product->discount->code ? 'selected' : '')?>
        >$discount->name</option>
<?php
endforeach;
?>
</select><br/>
<input type="hidden" name="code" value="<?=$this->
    product->code?>">
```

```
</form>
<button id="save">save</button>
<a href="<?=$this->url('inventory',['controller' =>
    'product'])?>">Go back</a>
```

The edit.phtml file will render the page shown in *Figure 7.6*:

Product insert

Name:

Price:

Discount:

save

Go back

© 2022 Laminas Project a Series of LF Projects, LLC.

Figure 7.6 – Product insert form

We have defined the content of our pages, but they don't work yet. In the following section, we will implement the interaction between the HTML pages and the product API.

Implementing the interaction between the web interface and API

We will use jQuery for implementing the REST interaction in the index.phtml and edit.phtml pages. The Laminas application skeleton already brings jQuery to the public/js folder. The layout.phtml file of the Application module not only imports the jQuery file but also the Bootstrap file too. However, these files are imported inline inside the HTML document body. We will change it for the files that are imported in the HTML document head:

1. Open the layout.phtml file in modules/Application/view/layout and remove the following part from the bottom of the file:

```
<?=$this->inlineScript()
    ->prependFile($this->basePath
```

```
                        ('js/bootstrap.min.js'))
            ->prependFile($this->basePath
                ('js/jquery-3.5.1.min.js')) ?>
```

Then, we will add the following part at the end of section <head> of layout.phtml.

```
        <?=$this->headScript()
            ->prependFile($this->basePath
                ('js/bootstrap.min.js'))
            ->prependFile($this->basePath
                ('js/jquery-3.5.1.min.js')) ?>
```

This change is necessary because we will add a JavaScript block to the document head, which depends on jQuery. We bring also Bootstrap to maintain uniformity.

2. Let us return to the Inventory module and modify the edit.phtml file in view/inventory/product with the following content:

edit.phtml

```php
<?php
$this->headScript()->appendScript
    (file_get_contents(__DIR__ . '/edit.js'));
?>
<h1>Product <?=(empty($this->product->code) ?
    'insert' : 'update')?></h1>
<form id="product">
    Name: <input type="text" name="name"
        autofocus="autofocus" value="<?=$this->
            product->name?>"><br/>
    Price: <input type="text" name="price"
        value="<?=$this->product->price?>"><br/>
    Discount: <select name="code_discount">
<?php
foreach($this->discounts as $discount):
?>
<option value="<?=$discount->code?>" <?=
    ($discount->code == $this->product->discount->code
    ? 'selected' : '')?> ><?=$discount->name?></option>
<?php
```

```
endforeach;
?>
</select><br/>
    <input type="hidden" name="code" value="<?=$this-
        >product->code?>">
</form>
<button id="save">save</button>
<p>
    <span id="api" url="<?=$this->url
        ('inventoryapi',['controller' =>
            'productapi'])?>"></span>
    <span id="target" url="<?=$this->url
        ('inventory',['controller' => 'product'])?>">
            </span>
    <a href="<?=$this->url('inventory',
        ['controller' => 'product'])?>">Go back</a>
</p>
```

Yes, the span tag does not have a url attribute, but it will work. We are doing it this way because it is easy to understand. There are more appropriate ways to pass data to JavaScript, but how JQuery is acting here is only a sample and is not required as a partner for Laminas, so we will accept this solution.

Let's see what has changed in the edit.phtml file since the last version:

- At the top of the file, we are telling the view layer to add a JavaScript file called edit.js to the document head. The edit.html file only produces the content for the HTML body. The complete HTML document is a combination of layout.html (which has the document head) with a view file of a controller.

- At the bottom of the file, we are defining two placeholders with a <span> tag for storing the URLs of the listing page and the product API.

Well, if edit.phtml reads an edit.js file, this last file needs to exist. Let us create this file in view/inventory/product with this source code:

edit.js

```
var save = function(){
    var api = $( "#api" ).attr("url");
    var target = $( "#target" ).attr("url");
    var formData = $( "#product" ).serialize();
```

```
    var code = parseInt($( "input[name='code']"
            ).val());
    if (code == 0){
        $.post( api, formData , function(data){
        alert("product has been inserted");
            window.location.href = target;
        }).fail(function(){
            alert('error');
        });
    } else {
        $.put( api, formData , function(data){
            alert("product has been upd
                        ated");
            window.location.href =  target;
        }).fail(function(){
            alert('error');
        });
    }
};
$(document).ready(function(){
    $("#save").click(save);
});
```

The edit.js file connects the OnClick event of the #save button with the save function. The save function reads the form and makes an HTTP request with the POST method if there is no code or with the PUT method if there is code. If the insertion occurs successfully, a window with confirmation opens and the user is redirected to the listing page.

Let us prepare the listing page for editing and deleting products. Change the index.phtml file in view/inventory/product to the following:

index.phtml

```php
<?php
$this->headScript()->appendScript(file_get_contents(__DIR__
    . '/index.js'));
?>
<h1>Product Registration</h1>
```

```php
<a href="<?=$this->url('inventory', ['controller' =>
    'product','action' => 'edit'])?>">Add product</a>
<table>
    <thead>
    <tr>
            <th>code</th>
            <th>name</th>
            <th>product</th>
            <th>discount</th>
    </tr>
    </thead>
    <tbody>
<?php
foreach ($this->products as $product) :
?>
    <tr>
        <td><button id="<?=$product->code?>"
class="edit"><?=$product->code?></button></td>
        <td><?=$product->name?></td>
        <td><?=$product->price?></
td>                    <td><?=$product->discount->
        name?></td>            <td><button id="<?=$product->code?>"
class="delete">delete</button></td>
        </tr>
<?php
endforeach;
?>
    </tbody>
</table>
<span id="api" url="<?=$this->url('inventoryapi',
    ['controller' => 'productapi'])?>"></span>
<span id="target" url="<?=$this->url('inventory',
    ['controller' => 'product'])?>"></span>
```

The following has changed in this file since the last version:

- At the top of the file, we are telling the view layer to add a JavaScript file called `index.js` to the document head. The `index.html` file only produces the content for the HTML body. As we have already said, the complete HTML document is a combination of `layout.html` (which has the document head) with a view file of a controller.

- The first column of rows is a button with a `class` attribute equal to `"edit"`.

- The last column of rows is a button with a `class` attribute equal to `"delete"`.

- At the bottom of the file, we are defining two placeholders with a `<span>` tag for storing the URLs of the listing page and the product API.

Well, if we are reading an `index.js` file, this file needs to exist. Let us create this file in `view/inventory/product` with this source code:

index.js

```
var edit = function(){
    var code = parseInt($(this).attr('id'));
    var target = $( "#target" ).attr("url") + '/edit/'
            + $(this).attr('id');
    if (code != 0){
            document.location.href = target;
    }
};
var remove = function(){
    var api = $( "#api" ).attr("url") + '/' +
            $(this).attr('id');
    If (confirm("Are you right to delete this
            product?")){
            $.ajax({url: api, type: 'DELETE'}).done
                    (function(data){
                alert("product has been
                            deleted");
                document.location.reload();
            }).fail(function(){
                alert('error');
            });
    }
```

```
};
$(document).ready(function(){
    $("button[class='edit']").click(edit);
    $("button[class='delete']").click(remove);
});
```

The index.js file connects the OnClick event for buttons with an edit class with an edit function. The index.js file also connects the OnClick event for buttons with a delete class with a remove function.

The edit function redirects us to the edit page with the product code as a parameter. In the remove function, if the user confirms the operation, an HTTP request with the DELETE method is made. If the deletion occurs successfully, a window is opened with the confirmation (see *Figure 7.7*) and the user is redirected to the listing page:

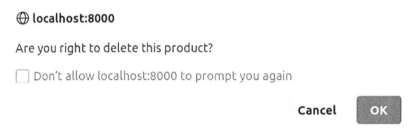

Figure 7.7 – Confirmation window for deleting a product

It seems that we have implemented the necessary tools for interacting with our pages and inserting, updating, deleting, and listing products. In fact, we need to make some adjustments to the view model. We will make them in the next section.

Testing the interaction between the web interface and API

For testing the product web pages, we need to adjust the Product and ProductTable classes. Why? Because of the relationship between products and discounts. To test the API, it was enough for the product model to have only the discount code, but to show the discount name on the product listing page, we need to recover the discount name.

Let us change the Product->exchangeArray method to this content:

```
public function exchangeArray($data): void
{
    $this->code = ($data['code'] ?? 0);
    $this->name = ($data['name'] ?? '');
```

```
$this->price = ($data['price'] ?? 0.0);
$this->discount = new Discount();
$this->discount->code = ($data['code_discount'] ??
    0);
$this->discount->name = ($data['name_discount'] ??
    '');
}
```

Observe that we have added a name_discount key.

Then, let us change the Product->toArray method to this content:

```
public function toArray()
{
    $attributes = get_object_vars($this);
    unset($attributes['discount']);
    unset($attributes['name_discount']);
    $attributes['code_discount'] = $this->
        discount->code;
    return $attributes;
}
```

Observe that we removed the name_discount key because this field is not persisted in the products table.

Where does this name_discount key come from? This key will be recovered by the ProductTable class, but we must override the getAll method in this class by making an INNER JOIN SQL statement. The ProductTable->getAll method must be as follows:

```
public function getAll($where = null): iterable {
    $select = new Select($this->tableGateway->
        getTable());
    $select->join('discounts', 'products.code_Discount=
        discounts.code', ['name_discount' => 'name']);
    if (!is_null($where)){
        $parsedWhere = [];
        foreach($where as $index => $value){
            $parsedWhere['products.' . $index] =
                $value;
        }
```

```
        $select->where($parsedWhere);
    }
    return $this->tableGateway->selectWith($select);
}
```

After these changes, we can test our web interface:

1. First, we will add a discount manually because a product depends on a discount. Open your terminal and go to the c:\xampp folder (in Windows) or the /opt/lampp folder (in Linux). Type one of the following commands to open the MySQL/MariaDB client:

   For Windows, see the following command:

   ```
   mysql\mysql -u root
   ```

   For Linux, see the following command:

   ```
   bin/mysql -u root
   ```

2. Select the whatstore database with the use command:

   ```
   MariaDB [(none)]> use whatstore;
   ```

3. We have created many codes with our tests. Let us test the product auto increment index using the ALTER TABLE command. The following code lines show the command and the expected output:

   ```
   MariaDB [whatstore]> ALTER TABLE products AUTO_INCREMENT
   = 1;
   Query OK, 0 rows affected (0.362 sec)
   Records: 0  Duplicates: 0  Warnings: 0
   ```

4. Restart the discount auto increment index too:

   ```
   MariaDB [whatstore]> ALTER TABLE discounts AUTO_INCREMENT
   = 1;
   Query OK, 0 rows affected (0.323 sec)
   Records: 0  Duplicates: 0  Warnings: 0
   ```

5. Add a discount:

   ```
   MariaDB [whatstore]> INSERT INTO discounts(name,
   operator, factor) VALUES ("no discount","-",0);
   Query OK, 1 row affected (0.152 sec)
   ```

Something useful – after you execute SQL INSERT, you can use the LAST_INSERT_ID() function to get this value:

```
MariaDB [whatstore]> SELECT LAST_INSERT_ID();
+------------------+
| LAST_INSERT_ID() |
+------------------+
|                1 |
+------------------+
1 row in set (0.000 sec)
```

6. Now, access http://localhost:8000/inventory/product/edit and fill the form with a product name, as per the example shown in *Figure 7.8*:

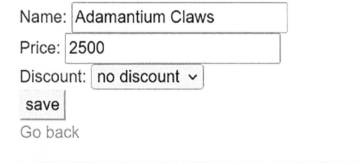

Figure 7.8 – A filled product insert form

7. Click on the **save** button. A window will be open with a **product has been inserted** message (see *Figure 7.9*):

⊕ **localhost:8000**

product has been inserted

OK

Figure 7.9 – Message window confirming the product insertion

Next, the user will be redirected to the product listing page (see *Figure 7.10*):

Product Registration

Add product

code	name	price	discount	
1	Adamantium Claws	2500	no discount	delete

© 2022 Laminas Project a Series of LF Projects, LLC.

Figure 7.10 – The product listing page after an insertion

8. You can try to edit the inserted product by clicking on the button with the product code. After that, you can try to delete the product by clicking on the **delete** button.

Finally, we have implemented CRUD with the Laminas controller and view layers. All the implementation for the `Inventory` module is available at `https://github.com/PacktPublishing/ PHP-Web-Development-with-Laminas/tree/main/chapter07/whatstore/ module/Inventory`.

Summary

In this chapter, we have learned how Laminas deals with HTTP requests and responses. We also understood how a module works across the Laminas request life cycle.

With exercises, we learned how to implement CRUD with a controller and view pages. First, we understood the relationship between HTTP and PHP. This introduction was necessary for us to understand the next topic – that is, the request life cycle in the Laminas MVC.

After establishing the foundations, we started our TDD approach to implementing CRUD for products. So, we checked the state of the existing tests. Next, we created new tests for inserting, recovering, updating, and deleting products.

Once we were finished with the tests for CRUD, we created a controller class for exposing an API for product management. Next, we created a controller class for the user interface, which works in tandem with the API controller class. For this last controller, we implemented two pages, one for product editing and another for product listing.

Finally, we implemented the interaction between the web interface and API, manually testing this interaction in sequence.

In the next chapter, we will evolve our application, using the `laminas-form` component to create web forms programmatically. In addition, we will also learn how to use filters and validators.

8

Creating Forms and Implementing Filters and Validators

This chapter will introduce some Laminas components to handle data input. In this chapter, you will learn how to connect the model layer with the view layer in a controller using `laminas-form`, how to filter input data with `laminas-filter`, and how to apply validation rules with `laminas-validator`.

In this chapter, we'll be covering the following topics:

- Generating a form for discounts with `laminas-form`
- Filtering input data
- Validating input data

At the end of this chapter, you will be able to define HTML forms programmatically using the `laminas-form` component. You will also know how to convert input data and remove unwanted (and possibly dangerous) content using the `laminas-filter` component. Finally, you will also discover how to validate input data before saving it into a database using the `laminas-validator` component.

Technical requirements

All the code related to this chapter can be found at `https://github.com/PacktPublishing/PHP-Web-Development-with-Laminas/tree/main/chapter08/whatstore`.

Generating a form for discounts with laminas-form

In this chapter, we will use three new Laminas components. The first is `laminas-form`, used to define HTML forms. In the same way you added and installed the `laminas-db` component using Composer, do the same for the `laminas-form` component:

1. Open the `composer.json` file in Eclipse.

2. Through the **Dependencies** tab, search for the `laminas/laminas-form` package and add it.

3. Save the file and click on the **Update dependencies** button.

    If you have any doubts about how to install Composer dependencies with Eclipse, read the *Creating our first automated test* section of *Chapter 3, Using Laminas as a Library with Test-Driven Development*.

Important note: Component injection into configuration

When you install a Laminas component in an MVC Laminas project with Composer, you will be asked about the injection of a component into configuration. When Composer asks this, you must select the option to inject into the `config/modules.config.php` file. We use this file because all Laminas components of our project must be available in the production environment. You can set Composer to remember this option when it asks about it.

Since the previous chapter, you are probably eager to create more APIs and web interfaces for other registrations of the `Inventory` module. But it was worth waiting until now because, in this section, you will create another **Create, Read, Update, Delete (CRUD)** implementation using `laminas-form` to define an HTML form for addition. In the next section, we will learn how to use `laminas-form` to implement the API and web interface for product discounts.

Creating CRUD implementation using the laminas-form component

Imagine you already have implemented CRUD for product discounts, in a similar way as we did in *Chapter 7, Request Control and Data View*, for products. It is not difficult to imagine because its source code is actually available in the repository mentioned in the *Technical requirements* section. As we are using common patterns, the structure of CRUD implementations is the same. So, we are assuming that there are the following classes in the `module/Inventory/src/Controller` folder:

*   `DiscountAPIController`
*   `DiscountAPIControllerFactory`
*   `DiscountController`
*   `DiscountControllerFactory`

You can see that we have two pairs of classes: an API controller and its factory and a web interface controller and its factory. It is exactly the same structure that we used for the product controller layer implementation. These controller classes are tested by the following classes in the `module/Inventory/test/Controller` folder:

- `DiscountAPIControllerTest`
- `DiscountControllerTest`

We also have a view layer implementation for discounts that is very similar to what we made for products. So, we are assuming that the following view scripts are in the `module/Inventory/view/inventory/discount` folder:

- `edit.js`
- `edit.phtml`
- `index.js`
- `index.phtml`

Here, we have two pairs of files that generate two web pages: the inserting/updating page and the listing page. The `.phtml` files have a combination of HTML and PHP code and the `.js` files have exclusively JavaScript code.

> **Important note: .phtml content**
> Although you can use PHP code in a `.phtml` file, it is a good practice to reduce this use. Model classes are responsible for business rules processing, so view script files must only show values sent from the models by controllers.

Right, imagine we have a discount registration working, as well as product registration. The `edit.phtml` file that generates an HTML form for inserting/updating has this content:

edit.phtml

```php
<?php
$this->headScript()->appendScript(file_get_contents
    (__DIR__ . '/edit.js'));
?>
    <h1>Discount <?= (empty($this->discount->code) ?
        'insert' : 'update')?></h1>
<form id="discount">
  Name: <input type="text" name="name"    autofocus=
```

```
    "autofocus" value="<?=$this->discount->name?>"><br/>
    Operator: <select name="operator">
    <option value="-" <?=($discount->operator == '-' ?
        'selected' : '')?> >-</option>
    <option value="*" <?=($discount->operator == '*' ?
        'selected' : '')?> >*</option>
    <option value="/" <?=($discount->operator == '/' ?
        'selected' : '')?> >/</option>
</select><br/>
    Factor: <input type="number" name="factor" value=
        "<?=$this->discount->factor?>"><br/>
    <input type="hidden" name="code" value="<?=$this->
        discount->code?>">
</form>
    <button id="save">save</button>
<p>
    <span id="api" url="<?=$this->url
        ('inventoryapi',['controller' => 'discountapi'])
            ?>"></span>
    <span id="target" url="<?=$this->url('inventory',
        ['controller' => 'discount'])?>"></span>
    <a href="<?=$this->url('inventory',['controller' =>
        'discount'])?>">Go back</a>
</p>
```

The source code of the edit.phtml file combined with the layout.phtml file (from the Application module) produces the web page shown in *Figure 8.1*. Observe that the content of edit.phtml is a simple HTML form whose elements have values dynamically defined by PHP code.

Discount insert

Name:

Operator: -

Factor: 0

save

Go back

© 2022 Laminas Project a Series of LF Projects, LLC.

Figure 8.1: Form for inserting discount

We will change the way this HTML form is generated. Of course, you could argue "Why do we need to define a form directly with HTML? There are JavaScript libraries that produce rich web interfaces. We could use them!" That's right, you could use any JavaScript library or framework for your user interface. Laminas does not require you to use HTML directly for creating forms or to use a JavaScript component. You are free to use what you wish for a frontend application. Remember we have APIs in our controller layer.

But it is good to know how to generate HTML forms programmatically from the application backend if you want to keep direct control over your web user interface. In addition, you will always have a default web interface to operate the system. More importantly, the Laminas component for generating forms is not only useful for generating a user interface but also for data validation. We will see this shortly.

We will delegate the generation of the discount form to `laminas-form`. To do that, we need to create a class to define the form.

Creating a class to define an HTML form

Follow these steps to create a class defining an HTML form:

1. First, in the `module/Inventory/src` folder, create a folder named `Form`.

2. Next, inside the `Form` folder, create a `DiscountForm.php` file. This file must contain a `DiscountForm` class:

DiscountForm.php

```php
<?php
namespace Inventory\Form;
use Laminas\Form\Form;
use Laminas\Form\Element\Text;
use Laminas\Form\Element\Select;
use Laminas\Form\Element\Number;
use Laminas\Form\Element\Hidden;
class DiscountForm extends Form {
    public function __construct($name = 'discount'){
        parent::__construct($name);
        $text = new Text('name');
        $text->setLabel('Name:');
        $text->setAttribute('autofocus', 'autofocus');
        $this->add($text);
        $select = new Select('operator');
        $select->setLabel('Operator:');
        $options = [
            '-' => '-',
            '*' => '*',
            '/' => '/',
        ];
        $select->setValueOptions($options);
        $this->add($select);
        $number = new Number('factor');
        $number->setLabel('Factor:');
        $this->add($number);
        $hidden = new Hidden('code');
```

```
            $this->add($hidden);
    }
}
```

We can observe the following from the DiscountForm class:

- DiscountForm inherits implementation from the Laminas\Form\Form class.
- The superclass constructor receives the value for the name attribute of the HTML form as an argument.
- Inside the inheritor Laminas\Form\Form class constructor, we define the form elements.
- Each HTML form element has a class with an appropriate name. For example, for HTML INPUT elements with type equal to "text," there is the Laminas\Form\Element\Text class.
- The Laminas\Form\Element class family groups HTML form elements.
- The Laminas\Form\Element constructor receives a value as the first argument, which will be used both for the id attribute and the name attribute.
- Both the Laminas\Form\Form and Laminas\Form\Element class constructors have a second argument, but it is optional.
- The setLabel method allows you to define a label for an HTML form element.
- The setAttribute method allows you to define the value of any HTML element attribute. For example, we have defined a value for the autotofocus attribute of the name input field.
- Laminas\Form\Element instances need to be associated to the Laminas\Form\Form instance by the Laminas\Form\Form->add method.

Right, but how do we use the DiscountForm class? Well, we need to create a DiscountForm instance in the web interface controller and pass it on to the appropriate view script. So, we will change the DiscountController->editAction method for the following:

```
    public function editAction()
    {
        $key = $this->params('key');
        $discount = $this->discountTable->getByField
            ('code',$key);
        $form = new DiscountForm();
        $form->bind($discount);
        return new ViewModel([
            'discount' => $discount,
            'form' => $form
```

```
        ]);
    }
```

Observe that we have created an instance of DiscountForm and we have invoked the DiscountForm->bind method. This method is inherited from Laminas\Form\Form. The bind method assigns the attribute values of a model object to the related fields of forms. We talk to DiscountForm to get the attribute values of the Discount model instance and use them to fill the HTML form elements. But how does the bind method extract the values from the model object? Well, the bind method invokes the getArrayCopy method from the model. Wait! The Discount class doesn't have this method. We need to implement it! In fact, it is better to implement the getArrayCopy method for all models, so every one of them can work with Laminas\Form\Form.

Because of this, we will add a getArrayCopy method to the AbstractModel class. This class was created in *Chapter 4, From Object-Relational Mapping to MVC Containers*, and it belongs to the Generic module. The implementation of the getArrayCopy method is simple:

```
    public function getArrayCopy()
    {
        return get_object_vars($this);
    }
```

Right. Now we can modify the edit.phtml file for discounts. The new implementation will be the following:

edit.pthml

```php
<?php
$this->headScript()->appendScript(file_get_contents
    (__DIR__ . '/edit.js'));
?>
<h1>Discount <?=(empty($this->discount->code) ? 'insert' :
    'update')?></h1>
<?php
$form = $this->form;
echo $this->form()->openTag($form);
echo $this->formLabel($form->get('name'));
echo $this->formText($form->get('name')) . '<br/>';
echo $this->formLabel($form->get('operator'));
echo $this->formSelect($form->get('operator')) . '<br/>';
echo $this->formLabel($form->get('factor'));
```

```
echo $this->formNumber($form->get('factor')) . '<br/>';
echo $this->formHidden($form->get('code'));
echo $this->form()->closeTag();
?>
<button id="save">save</button>
<p>
    <span id="api" url="<?=$this->url('inventoryapi',
        ['controller' => 'discountapi'])?>"></span>
    <span id="target" url="<?=$this->url('inventory',
        ['controller' => 'discount'])?>"></span>
    <a href="<?=$this->url('inventory',['controller' =>
        'discount'])?>">Go back</a>
</p>
```

Observe that we have replaced the content of the HTML form element with a sequence of calls for methods prefixed with `form`. We need to understand the following from the previous source code:

- The `$this->form` attribute is the `DiscountForm` instance sent by the `DiscountController` instance.

- The `$this->form` method (with parentheses at the end) is a view helper that renders HTML forms defined with `Laminas\Form\Form`.

- The `$this->formLabel` method renders only the label of a form element.

- Each element type has an appropriate render helper class. For example, the operator field is an HTML SELECT element, so we use `$this->formSelect` to render it.

- The `get` method of `Laminas\Form\Form` returns a `Laminas\Form\Element` object. The argument method is the value passed into the object constructor.

To use the view helper classes, we need to have `Laminas\Form` as a declared module in the `modules.config.php` file. If there is no element with the `"Laminas\Form"` value in the array, which is returned by `modules.config.php`, you will receive a message such as **A plugin by the name 'form' was not found in the plugin manager Laminas\View\HelperPluginManager**, when trying to render the form.

Adding `Laminas\Form` as a module is a concise way to enable form rendering. There is a more verbose way to do it. We will go through it here because it will help you to understand what happens behind the scenes, that is, what `laminas-mvc` does for you.

The verbose way to enable form rendering is by registering aliases and factories for each one of the form helpers in the `module.config.php` file. This register is made by the key `view_helpers`, at the first level of the configuration array, as we can see in this source code:

```
'view_helpers' => [
    'aliases' => [
        'form' => Form::class,
        'formText' => FormText::class,
        'formSelect' => FormSelect::class,
        'formNumber' => FormNumber::class,
        'formHidden' => FormHidden::class,
        'formLabel' => FormLabel::class
    ],
    'factories' => [
        Form::class => InvokableFactory::class,
        FormText::class => InvokableFactory::class,
        FormSelect::class => InvokableFactory::class,
        FormNumber::class => InvokableFactory::class,
        FormHidden::class => InvokableFactory::class,
        FormLabel::class => InvokableFactory::class
    ]
]
```

In the `aliases` subkey, we defined the names in which we invoked view helper classes. In the `factories` subkey, we defined how the view helper instances are created. Observe that this approach allows us to make customizations. This is often a feature of Laminas components.

You will have noticed that it is better to add `Laminas\Form` as a module. This is what we will do in our project.

> **Important note: Form view plugins**
>
> If you haven't declared `Laminas\Form` as a module and you haven't registered a form view plugin in `module.config.php`, you will receive a message such as **A plugin by the name 'form' was not found in the plugin manager Laminas\View\HelperPluginManager** when trying to render a form.

We also need to install the `laminas-i18n` component before trying to render the form. This component allows you to translate text from translation files according to the application configuration. Install `laminas-i18n` as you have done with `laminas-form`.

> **i18n**
>
> i18n is the acronym for internationalization. In short, it is support for various languages. I18n software components help to translate content from a language to another one.

> **Important note: Dependency between a form and translator**
>
> If you haven't installed the `laminas-i18n` component and you try to render a form defined by `laminas-form`, you will receive the message **Class 'Laminas\I18n\View\Helper\AbstractTranslatorHelper' not found**. Well, according to this message, a form requires a translator. This requirement can seem strange because it supposes that any form must be translated. Considering a hard coupling architecture, such as the Laminas components architecture, the relationship between a form component and a translator component should be optional. But the reason for the strong coupling between `laminas-form` and `laminas-i18n` is that form components can be integrated into filter and validation components and some of these components are dependent on `laminas-i18n`.

> **laminas-mvc-form**
>
> If you are using `laminas-mvc` and you don't want to install `laminas-form` and `laminas-i18n` separately, you can install the `laminas-mvc-form` meta package.

At this time, we have successfully replaced the direct definition of the discount form with HTML by rendering `laminas-form`. The presentation of the discount form has not changed, so we don't have a new interface. On the other hand, we can now connect the model to the form in the controller layer.

So, we have concluded the generation of a form for adding and updating discounts using `laminas-form`. To do this, we did the following:

- Created a CRUD implementation for discounts, with an API and a web interface
- Created a class to define an HTML form for discounts

In the next section, we will learn how to filter input values using `laminas-filter`.

Filtering input data

As you have added and installed the `laminas-form` component in this chapter using Composer, do the same for the `laminas-inputfilter` component:

1. Open the `composer.json` file in Eclipse.
2. Through the **Dependencies** tab, search for the `laminas/laminas-inputfilter` package and add it.
3. Don't forget to save the file and click on the **Update dependencies** button.

The laminas-inputfilter component combines laminas-filter and laminas-validator. You can use laminas-filter and laminas-validator separately in PHP programs, but the use of laminas-inputfilter is more convenient for MVC web applications.

We have an HTML form to insert/update discounts. But this is one of the ways to make requests for handling discounts. We have an API, so another application can make requests to *whatstore*. Our application is oriented to service, so it is open to multiple clients. This is useful, but there is an issue with the data handling. If a user fills in an HTML form, JavaScript code can filter form data before sending the request. When a request is made without an HTML page, there is no JavaScript to filter the data. Then, the server must ensure a valid format for input data.

Of course, the valid format for application data depends on application requirements. We are assuming here that we are reading requirement documents for implementing the code for handling data.

The place to define the filtering rules for the data of a model is the model class. So, we will add a getInputFilter method to the Discount class, according to the following source code:

```php
public function getInputFilter(): InputFilter
{
    $inputFilter = new InputFilter();
    $input = new Input('code');
    $filterChain = new FilterChain();
    $filterChain->attach(new ToInt());
    $input->setFilterChain($filterChain);
    $inputFilter->add($input);
    $input = new Input('name');
    $filterChain = new FilterChain();
    $filterChain->attach(new StringToUpper())
    ->attach(new Alnum(true));
    $input->setFilterChain($filterChain);
    $inputFilter->add($input);
    $input = new Input('operator');
    $filterChain = new FilterChain();
    $filterChain->attach(new AllowList(['list' =>
        ['-','*','/']]));
    $input->setFilterChain($filterChain);
    $inputFilter->add($input);
    $input = new Input('factor');
    $filterChain = new FilterChain();
    $filterChain->attach(new ToFloat());
```

```
        $input->setFilterChain($filterChain);
        $inputFilter->add($input);
        return $inputFilter;
    }
```

The `InputFilter` class is a collection of `Input` instances. The `Input` class allows you to chain filters and validators for an input attribute. For each `Discount` class attribute, we have created an instance of `Input`.

To connect filters to an input attribute, we use the `FilterChain` class. `FilterChain` connects a group of filters and allows you to apply all these filters to a value. In sequence, these filters are attached to `FilterChain`.

You can find a reference about Laminas standard filter classes at `https://docs.laminas.dev/laminas-filter/standard-filters`. For the `Discount` class, we have used the following filter classes:

- The `ToInt` class was used to ensure that discount code is integer.

- The `StringToUpper` class was used to ensure capital letters for the discount name. In addition, `Alnum` ensures that only alphanumeric values are allowed. In fact, the argument `true` in the `Alnum` constructor allows whitespaces too.

- The `AllowList` class was used to restrict the possible values of the operator to three possibilities.

- The `ToFloat` class was used to ensure that factor is a float.

> **Important note: Filter classes**
>
> There are more available filters other than standard filters. You can find a complete list of available Laminas filters at `https://docs.laminas.dev/laminas-filter`. In addition, you can create your own filter classes by implementing `Laminas\Filter\FilterInterface`. We will learn how to create customized filters in *Chapter 13, Tips and Tricks*.

Since we have defined a method for generating an instance of `InputFilter`, we will use this instance for transforming the attribute values of the `Discount` model. It is done with the `Discount->toArray` method, like so:

```
    public function toArray()
    {
        $inputFilter = $this->getInputFilter();
        $inputFilter->setData(get_object_vars($this));
        return $inputFilter->getValues();
    }
```

This change is enough to ensure the application of filters. You can now insert or update a discount, and you will see that the name will be listed with capital letters, as you can see in *Figure 8.2*:

Figure 8.2: Effect of the StringToUpper filter

In the next section, we will learn how to validate input values using `laminas-validator`.

Validating input data

Filters transform data, but they are not enough to ensure that data is appropriate for our system. Filters just perform an action. To know whether data has appropriate values, we need a component that answers the question does the data comply with the business rule?

Laminas has a component for this: `laminas-validator`. As we've already said, `laminas-validator` is a dependency of `laminas-inputfilter`. Because of this, you don't need to install `laminas-validator` if you already have installed `laminas-inputfilter`.

We said that validation depends on business rules. Imagine your client told you when they were asked about the requirements that discount names must have a minimum of three characters. We have a filter to change alphabetic characters to capital letters and another filter to remove characters that are not alphabetic or numeric or whitespaces. However, these filters do not ensure that the final name has at least three characters.

We need to create a validation rule for the name field. Let us modify the part of the `Discount->getInputFilter` method that handles the name field:

```
$input = new Input('name');
$filterChain = new FilterChain();
$filterChain->attach(new StringToUpper())
->attach(new Alnum(true));
$input->setFilterChain($filterChain);
```

```
$validatorChain = new ValidatorChain();
$validatorChain->attach(new StringLength
    (['min' => 3]));
$input->setValidatorChain($validatorChain);
$inputFilter->add($input);
```

Observe that we have created an instance of `ValidatorChain`. `ValidatorChain` connects a group of validators and allows you to apply all these validators to a value. In sequence, these validators are attached to `ValidatorChain`.

You can find the reference about Laminas standard filter classes at `https://docs.laminas.dev/laminas-validator/standard-validators`. For the `Discount` class, we have used the `StringLength` validator. This validator ensures that a text has a minimum and/or maximum length.

We will add validation in the inserting and updating methods of `DiscountAPIController`. `DiscountAPIController->create` must look like this:

```
public function create($data)
{
    $discount = new Discount();
    $inputFilter = $discount->getInputFilter();
    $inputFilter->setData($data);
    if (!$inputFilter->isValid()){
        return new JsonModel(['inserted' => 'invalid']
            );
    }
    $discount->exchangeArray($data);
    $inserted = $this->discountTable->save($discount);
    return new JsonModel(['inserted' => $inserted]);
}
```

Observe that we applied the validation rule for the discount name before trying to save the data. The database could have business rules too, but validation in the application avoids database errors and unnecessary communication between the application and the database. We get the `InputFilter` instance with the `Discount->getInputFilter` method, which makes sense. The `InputFilter->setData` method assigns the values to be validated and the `InputFilter->isValid` method answers if all values are valid according to the validator chain.

We will also modify `DiscountAPIController->update`:

```
public function update($id, $data)
{
    $discount = $this->discountTable->getByField
        ('code', $id);
    $inputFilter = $discount->getInputFilter();
    $inputFilter->setData($data);
    if (!$inputFilter->isValid()){
        return new JsonModel(['inserted' => 'invalid']
            );
    }
    $discount->exchangeArray($data);
    $updated = $this->discountTable->save($discount);
    return new JsonModel(['updated' => $updated]);
}
```

You can observe that if data is not valid, the `action` method returns an `inserted` key with the string value `invalid`. We will use this value to show an appropriate message to the user and avoid unnecessary redirecting. In the following code snippets, we show the content of the `edit.js` file for discounts. First, we will declare a `save` function. This function starts with the declaration of variables, which receives values from the page `edit.phtml`:

```
var save = function(){
    var api = $( "#api" ).attr("url");
    var target = $( "#target" ).attr("url");
    var formData = $( "#discount" ).serialize();
    var code = parseInt($( "input[name='code']" ).val());
```

The value of the code variable indicates whether the operation is an insert or an update. For an insert, we send the form data through an HTTP POST request:

```
if (code == 0){
    $.post( api, formData , function(data){
        if (data.inserted == 'invalid'){
            alert('Invalid data!');
            return;
        }
        alert("discount has been inserted");
```

```
                window.location.href = target;
        }).fail(function(){
                alert('error');
        });
```

For an update, we send the form data through an HTTP PUT request:

```
    } else {
            api = $( "#api" ).attr("url") + '/' + $(this).
                attr('code');
            $.ajax({url: api, type: 'PUT', data: formData}).
                done(function(data){
                    if (data.inserted == 'invalid'){
                            alert('Invalid data!');
                            return;
                    }
                    alert("discount has been updated");
                    window.location.href =  target;
            }).fail(function(){
                    alert('error');
            });
    }
};
```

Finally, the save function is associated with the button click event:

```
$(document).ready(function(){
        $("#save").click(save);
});
```

The modification focuses on the testing of asynchronous request content. If when returned, the inserted key has an invalid value, a message is shown, explaining that data is not valid. So, if you try to add a discount with a name shorter than three characters, you will see a message like in *Figure 8.3*:

Figure 8.3: Message for invalid data in inserting or updating actions

> **Important note: Validator classes**
>
> There are more available validators than standard validators. You can find a complete list of available Laminas validators at https://docs.laminas.dev/laminas-validator. In addition, you can create your own filter classes, extending Laminas\Validator\ AbstractValidator. We will learn how to create customized filters in *Chapter 13, Tips and Tricks*.

So, we have now finished this chapter, after modifying the generation of the HTML form and adding filters and a validator for our discount registration.

Summary

In this chapter, we have learned about some Laminas components related to data input. First, we learned how to define HTML forms programmatically using laminas-form and how to render such forms in .phtml files. We defined an HTML form for discount registration, taking the example of laminas-form.

Next, we learned how to filter values from requests with laminas-filter. We applied filters to discount fields, restricting, for example, the discount names to being alphanumeric with whitespaces.

Finally, we learned how to validate values from requests with laminas-validator. We applied a validator to the name field, avoiding saving names shorter than three characters.

Filtering and validating are aspects related to application security. A lack of control over data input can allow unwanted and sometimes dangerous content. Imagine a part of an input that is used as a parameter of a command invoked by an application. If anything at all is allowed, you don't know what might be executed by your application.

In the next chapter, we will talk about other aspects of security, that is, authentication and authorization.

9

Event-Driven Authentication

This chapter covers one of the essential topics for developing secure applications – authentication. Authentication is an identity check. An example of authentication in the physical world is the request for an identity document to get into a building.

In this chapter, you will learn how to implement this checking process with event-oriented programming. We will first create tests for login and logout actions, and then we will create these actions. After that, we will implement user authentication for the `Inventory` module. For this, we will create an employee registration. Finally, we will implement a user identity check based on events.

In this chapter, we'll be covering the following topics:

- Creating login and logout actions for the `Inventory` module
- Implementing user authentication for the `Inventory` module
- Creating employee registration
- Verifying the identity of users based on events

At the end of this chapter, you will be able to control who can access restricted areas of a web application.

Let us first start creating a login page in the `Inventory` module and the related controller action.

Technical requirements

All the code related to this chapter can be found at https://github.com/PacktPublishing/PHP-Web-Development-with-Laminas/tree/main/chapter09/whatstore.

Creating login and logout actions for the Inventory module

Continuing our **test-driven development** (TDD) approach, in this section, we will first create tests for login and logout actions, and then we will create form and controller classes and view scripts to be tested.

Creating tests for login and logout actions

As we have learned, when we create tests, we are writing requirements as source code. Tests say what we are expecting classes to do. Now, we will add two test methods in the `IndexControllerTest` class of the `Inventory` module – one method for testing the login action and another method for testing the logout action. The source code of the login test method looks like this:

```
public function testLoginActionCanBeAccessed(): void
{
    $_POST = [
        'nickname' => 'Jonn Jonnz',
        'password' => 'hronmeer'
    ];
    $this->dispatch('/inventory/index/login', 'POST',
        $_POST);
    $this->assertResponseStatusCode(302);
    $this->assertModuleName('inventory');
    $this->assertControllerName('index');
    // as specified in router's controller name alias
    $this->assertControllerClass('IndexController');
    $this->assertMatchedRouteName('inventory');
}
```

The source code of the logout test method looks like this:

```
public function testLogoutActionCanBeAccessed(): void
{
    $this->dispatch('/inventory/index/logout', 'GET');
    $this->assertResponseStatusCode(200);
    $this->assertModuleName('inventory');
    $this->assertControllerName('index');
```

```
            // as specified in router's controller name alias
            $this->assertControllerClass('IndexController');
            $this->assertMatchedRouteName('inventory');
    }
```

Now, maybe you're asking, "But how will the `testLoginActionCanBeAccessed` method ensure that a user can log in if this user does not exist?" At first, it does not ensure this. If you look closely, the `testLoginActionCanBeAccessed` method checks whether the HTTP response has the `302` code – in other words, the `loginAction` method – redirected to the other URL. If the user exists, they will be redirected to the menu page. The test will fail at this moment, as often happens with tests in TDD. But now, after having tests that fail, we can build our new actions.

Creating login and logout actions

If you access `http://localhost:8000/inventory` now, you will see the splash page of Laminas because the `Inventory` homepage was copied from the `Application` module:

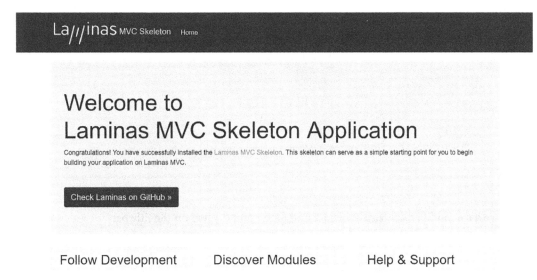

Figure 9.1 – The homepage of the Inventory module

We will modify this page. First of all, we will modify the page layout. According to what we said in *Chapter 4*, *From Object-Relational Mapping to MVC Containers*, page layout is defined by the `Application` module. We will change some parts of the `layout.phtml` file in the `module/Application/view/layout` folder.

The first part of the layout file to be modified is the title definition. We define the title directly with the TITLE tag, but Laminas has a view helper for this. You must modify the call for $this->headTitle for it to look like this:

```
<?= $this->headTitle('Whatstore')->setSeparator(' - ')->
    setAutoEscape(false) ?>
```

You may ask, "Why do I need to use a method only to write text that won't change?" In fact, when you use the $this->headTitle method for generating the TITLE tag instead of writing it directly, you allow the view scripts to easily modify the title.

Another part to be modified is the A tag, with a class attribute equal to navbar-brand. This is the tag that shows the Laminas logo and the **MVC Skeleton** text. The new content for this tag is the following:

```
<a class="navbar-brand" href="<?= $this->url('home') ?>">
Whatstore
</a>
```

The last part of the layout file to be modified is the FOOTER tag. The new content for this tag is the following:

```
<footer>
    <p>
        &copy; <?= date('Y') ?>
        Powered by Sheat Software
    </p>
</footer>
```

In short, we have modified the visible header and footer and the text on the title bar.

Now, we will create a class to define the login form. In the module/Inventory/src/Form folder, create a file named LoginForm.php, and add the following code:

LoginForm.php

```php
<?php
namespace Inventory\Form;

use Laminas\Form\Form;
use Laminas\Form\Element\Text;
```

```php
use Laminas\Form\Element\Password;

class LoginForm extends Form {
    public function __construct($name = 'user'){
        parent::__construct($name);
        $text = new Text('nickname');
        $text->setLabel('NickName:');
        $text->setAttribute('autofocus', 'autofocus');
        $this->add($text);
        $password = new Password('password');
        $password->setLabel('Password:');
        $this->add($password);
    }
}
```

As you can see, the `LoginForm` class is an extension of `Laminas\Form\Form` and uses element classes that we already used in *Chapter 8, Creating Forms and Implementing Filters and Validators*. We will use `LoginForm` in the `IndexController` class of the `Inventory` module. Modify the `IndexController->indexAction` method like so:

```php
    public function indexAction()
    {
        $form = new LoginForm();
        return new ViewModel([
            'form' => $form
        ]);
    }
```

The `$form` object is delivered to the view layer, and it will be handled by the `index.phtml` file in `modules/Inventory/view/inventory/index`. The content of `index.phtml` looks like this:

index.phtml

```php
<?php
/**
 * @var Laminas\View\Renderer\PhpRenderer $this
 */
$this->headScript()->appendScript(file_get_contents
    (__DIR__ . '/index.js'));
```

```php
?>
<h1>Inventory Module</h1>
<?php
$form = $this->form;
echo $this->form()->openTag($form);
echo $this->formLabel($form->get('nickname'));
echo $this->formText($form->get('nickname')) . '<br/>';
echo $this->formLabel($form->get('password'));
echo $this->formPassword($form->get('password'));
echo $this->form()->closeTag();
?>
<button id="login">login</button>
<p>
<span id="api" url="<?=$this->url('inventory',
        ['controller' => 'index', 'action' =>'login'])?>">
            </span>
<span id="target" url="<?=$this->url('inventory',
        ['controller' => 'menu'])?>"></span>
<a href="<?=$this->url('inventory')?>">Go back</a>
</p>
```

Observe that `index.phtml` imports an `index.js` file. The `index.js` file is responsible for submitting HTTP requests to the login API. Let us see the source code of `index.js`:

```javascript
var login = function() {
  var api = $("#api").attr("url");
  var target = $("#target").attr("url");
  var formData = $("#user").serialize();
  $.post(api, formData, function(data) {
    if (!data.logged) {
      alert('Invalid data!');
      return;
    }
    alert("user has been logged successfully");
    window.location.href = target;
  }).fail(function() {
    alert('error');
```

```
    });
};

$(document).ready(function() {
    $("#login").click(login);
});
```

You can see that `index.js` associates the `login` function with the event of the button click. The `login` function sends an HTTP POST request with data from the login form.

At this point, the tests fail yet, because the `loginAction` and `logoutAction` methods do not exist in the `IndexController` class. So, let us create them:

```
public function loginAction()
{
    $nickname = $this->request->getPost('nickname');
    $password = $this->request->getPost('password');

    $logged = false;

    return new JsonModel([
        'logged' => $logged
    ]);
}
public function logoutAction()
{
    return $this->redirect()->toRoute('inventory');
}
```

At this point, the `loginAction` method is incomplete. Because of this, the tests fail yet again. You will receive an HTTP 404 error for each one of the new `IndexController` methods. For example, for the `loginAction` method test, the output of PHPUnit must be this:

```
Failed asserting response code "302", actual status code is
"404"
```

If you request `http://localhost:8000/inventory/index/login` in the browser, you will see the page seen in *Figure 9.2*:

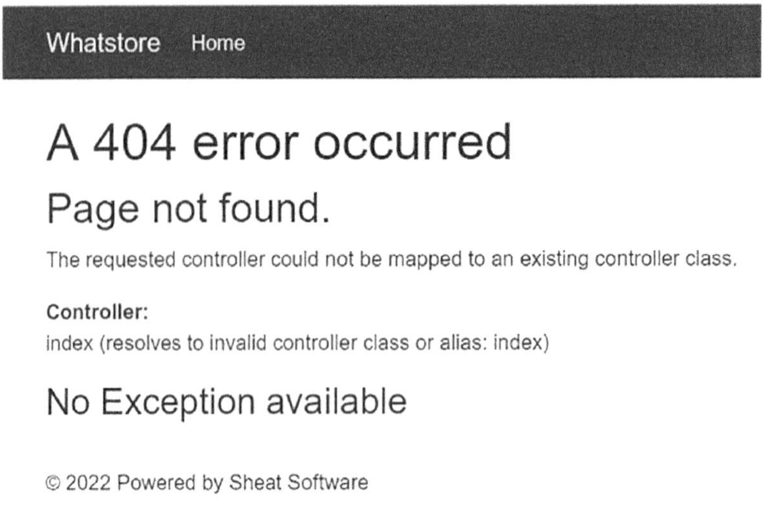

Figure 9.2 – Error page for controller not found

Why are we receiving this error? Because there is no alias for `IndexController` class in the `Inventory` module. If you observe the part of the test that checks the controller name, you can see the following comment about the alias:

```
$this->assertControllerName('index');
    // as specified in router's controller name alias
```

Let us solve this issue. In the `module.config.php` file, add a line that associates the `index` alias with `IndexController` class like so:

```
        'controllers' => [
            'aliases' => [
                'index' => Controller\IndexController::class,
```

This change will reduce the failures to only one, related to the `loginAction` method of `IndexController` class. In the next section, we will implement user authentication to solve this failure.

Implementing the user authentication for the Inventory module

For authenticating the user, we will use the `laminas-authentication` component. Add and install the `laminas-authentication` component in the same way you did with the laminas-db component, using Composer. Open the composer.json file in Eclipse, and through the **Dependencies** tab, search for the laminas/laminas-authentication package and add it. Don't forget to save the file and click on the **Update dependencies** button. If you have any doubt about how to install Composer dependencies with Eclipse, read the *Creating our first automated test* section in *Chapter 3, Using Laminas as a Library with Test-Driven Development*.

After installing laminas-authentication, we can use this component to modify the IndexController->loginAction method. This method will look like this:

```php
public function loginAction()
{
    $nickname = $this->request->getPost('nickname');
    $password = $this->request->getPost('password');
    $this->adapter->setIdentity($nickname);
    $this->adapter->setCredential($password);

    $logged = false;

    $auth = new AuthenticationService();
    $auth->setAdapter($this->adapter);

    $result = $auth->authenticate();

    if ($result->isValid()) {
        error_log('user ' . $result->getIdentity() . '
            is logged');
        $logged = true;
    } else {
        foreach ($result->getMessages() as $message) {
            error_log($message);
        }
    }

    return new JsonModel([
        'logged' => $logged
    ]);
}
```

We have made the following changes in the `loginAction` method:

- We have passed in the `$nickname` variable to the `setIdentity` method of the `$adapter` attribute.

- We have passed in the `$password` variable to the `setCredential` method of the `$this->adapter` attribute.

- We have instantiated the `Laminas\Authentication\AuthenticationService` class.

- We have passed in the `$this->adapter` attribute to the `setAdapter` method of the `Laminas\Authentication\AuthenticationService` instance.

- We have invoked the `authenticate` method from the `Laminas\Authentication\AuthenticationService` instance, and we have assigned the returned value to the `$result` variable.

- We have checked whether the `isValid` method of the `$result` object is `true`. If it is `true`, we send a message to the log and change the state of `$logged` to `true`.

- If `$result` is `false`, we send each generated error message to the log.

Wait! Where does `$this->adapter` attribute come from? Good question. We need to declare this attribute like so:

```
private ?AdapterInterface $adapter;
```

This `$adapter` attribute is `Laminas\Authentication\Adapter\AdapterInterface`.

Important note – namespaces

We already have used `AdapterInterface` for database, but it was from another namespace. Remember that there can be several classes with the same name for different namespaces. **Adapter** is a design pattern, and it can be applied to several classes. So, we have more than an `Adapter` class available in Laminas.

Okay, we need to assign the `$adapter` attribute to an instance that implements `Laminas\Authentication\Adapter\AdapterInterface`. We will do this in the constructor method:

```
public function __construct(AdapterInterface $adapter)
{
    $this->adapter = $adapter;
}
```

Everything could be right if we instantiated the controllers directly, but we don't instantiate controllers in Laminas. Laminas instantiates controllers for us through factories, which we talked about in *Chapter 1, Introducing Laminas for PHP Applications*. It is like the famous phrase of failed Hollywood auditions: *don't call us, we'll call you*. Therefore, we must create a factory class named `IndexControllerFactory` with this source code:

```php
<?php
namespace Inventory\Controller;

use Laminas\ServiceManager\Factory\FactoryInterface;
use Interop\Container\ContainerInterface;
use Laminas\Authentication\Adapter\DbTable\
    CredentialTreatmentAdapter;

class IndexControllerFactory implements FactoryInterface
{

    public function __invoke(ContainerInterface $container,
        $requestedName, array $options = null)
    {

        $employeeTable = $container->get('EmployeeTable');
        $adapter = new CredentialTreatmentAdapter
            ($container->get('DbAdapter'));
        $adapter->setTableName('employees');
        $adapter->setIdentityColumn('nickname');
        $adapter->setCredentialColumn('password');

        return new IndexController($adapter);
    }

}
```

This factory class instantiates the `CredentialTreatmentAdapter` class. The `CredentialTreatmentAdapter` class allows us to authenticate users with a database table. It requires an injection of a database adapter, since it is necessary to access a table. Once the table name, identity column (user name), and credential column (user password) are configured, the `IndexControllerFactory` class injects a `CredentialTreatmentAdapter` instance into the `IndexController` instance.

Okay, is it possible to authenticate now? Let's try. Access `http://localhost:8000/inventory`, and you will see the page shown in *Figure 9.3*:

Figure 9.3 – The login page for Inventory Module

Try to fill the fields with random values and click on the **login** button. The result will be a window with the **error** message, as seen in *Figure 9.4*:

Figure 9.4 – The error message after the login attempt

So, what is wrong? If you check the HTTP response through the browser debugger, you will the following error message: `Class "Laminas\Session\Container" not found`. If you run the tests, you also will see this message:

```
1) InventoryTest\Controller\
IndexControllerTest::testLoginActionCanBeAccessed
```

```
Failed asserting response code "302", actual status code is
"500"
```

```
Exceptions raised:
Exception 'Error' with message 'Class "Laminas\Session\
Container" not found' in /opt/lampp/htdocs/whatstore/vendor/
laminas/laminas-authentication/src/Storage/Session.php:62
```

This error occurs in a class of `laminas-authentication`. But, wait! If `laminas-authentication` needs this `Container` class, why didn't Composer install it? It is because Laminas components try to keep a weak coupling among them. You are not required to use `Container` to store authentication data. You can use another class. It is possible to inject another storage class in the `AuthenticationService` constructor. However, as we don't use a customized storage class, `laminas-authentication` requires one and asks for implementation available at another component – `laminas-session`, by default.

No problem. We will use `laminas-session` in the next chapter too. Add and install the `laminas-session` component, similar to how we did so with the `laminas-authentication` component using Composer in this chapter. Open the `composer.json` file in Eclipse, and through the **Dependencies** tab, search for the `laminas/laminas-session` package and add it. Don't forget to save the file and click on the **Update dependencies** buttons. Are you forgetting to save the file again, right? Don't worry, we will repeat as many times as it is necessary.

After installing `laminas-session`, you can try logging in again by the web interface. The result will be a window with the message **Invalid data!**, as seen in *Figure 9.5*:

Figure 9.5 – A warning message after a login attempt

Of course, at this moment, whichever nickname and password you type are invalid data, because there are no registered employees. For that, we need to create the employee registration pages. This is what we will do in the next section.

Creating the employee's registration

Maybe we offered a false promise in the last section. In fact, imagine you already have implemented a CRUD operation for employees, similar to what we did in *Chapter 7*, *Request Control and Data View*, for products. Also imagine that we have implemented this CRUD operation using `laminas-form`, `laminas-filter`, and `laminas-validation`, similar to what we did in *Chapter 8*, *Creating Forms and Implementing Filters and Validators*, for product discounts. It is not difficult to imagine because its source code is actually available in the repository mentioned in the *Technical requirements* section. As we are using patterns, the structure of CRUD implementations is the same. Therefore, we are assuming that there are the following classes in the `module/Inventory/src/Controller` folder:

- `EmployeeAPIController`
- `EmployeeAPIControllerFactory`
- `EmployeeController`
- `EmployeeControllerFactory`

Note that we have two pairs of classes – an API controller and its factory, and a web interface controller and its factory. It is exactly the same structure that we have used for product controller layer implementation. These controller classes are tested by the following classes in the `module/Inventory/test/Controller` folder:

- `EmployeeAPIControllerTest`
- `EmployeeControllerTest`

We also have a view layer implementation for discounts that is very similar to what we had made for products. Therefore, we are assuming that the following view scripts are in the `module/Inventory/view/inventory/employee` folder:

- `edit.js`
- `edit.phtml`
- `index.js`
- `index.phtml`

Here, we have two pair of files that generates two web pages – the inserting/updating page and the listing page. The `.phtml` files have a combination of HTML and PHP code, and the `.js` files have JavaScript code, exclusively.

Right, now we can register employees. We can insert an employee and check whether they can log in and log out. We will do it by following these steps:

1. First, access `http://localhost:8000/inventory/employee`. This URL will lead you to the page shown in *Figure 9.6*:

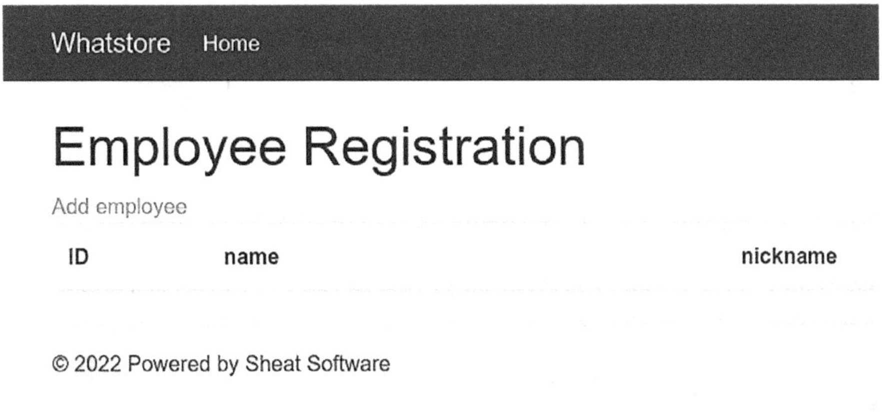

Figure 9.6 – The listing page of employees

2. Click on the **Add employee** option. This action will take you to the page shown in *Figure 9.7*. In this example, we are adding an employee named Steven Grant, whose nickname is Mark Spektor, and the password is "`moonknight`".

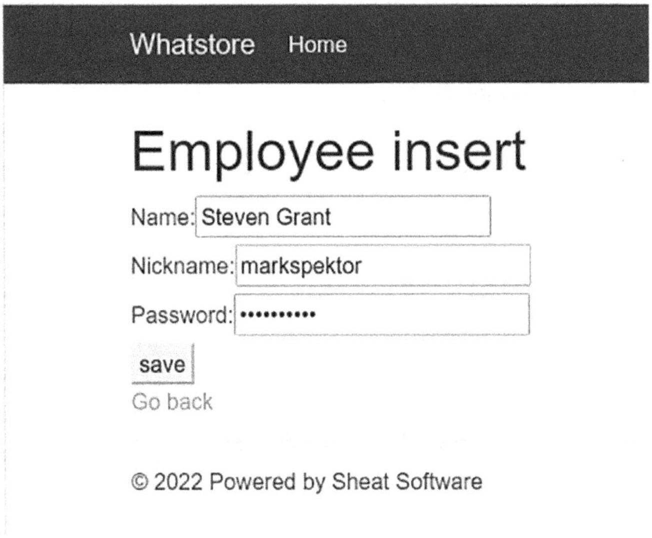

Figure 9.7 – The insertion form of employees

3. After clicking the **Save** button, you will see a message confirming the insertion, as we can see in *Figure 9.8*:

Figure 9.8 – The confirmation message for employee insertion

Finally, you will see the inserted employee on the listing page. Observe that filters have converted the letters to capitals, as *Figure 9.9* shows:

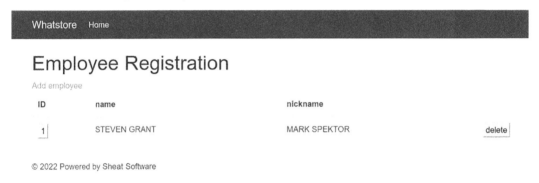

Figure 9.9 – The employee listing page with the inserted employee

Great! Now we have a user for authenticating. Let us use this user to test our login page in the next section.

Testing the login page

According to the example, we access `http://localhost:8000/inventory` and fill the login form with the data of the inserted employee.

You might get frustrated when, after clicking on the **login** button, a message saying **Invalid data!** appears. "But what is wrong?" you might think. You must have typed the employee nickname and password correctly, right? I believe you. In fact, there is a little detail related to the password – it is encrypted when the employee is inserted. So, we have to compare the encrypted password provided by the employee with the encrypted password stored in the `employees` table.

This comparison requires some changes in `Employee` class. You should remember that the `Employee` class has a private `encrypt` method. We will modify this method to `public` and `static`, like so:

```
public static function encrypt(String $text)
{
    $text = strrev(hash('sha256', $text));
    $subtext = substr($text,1,rand(0,strlen($text)-1));
    $text = substr($subtext . $text, 0, strlen($text));
    return hash('md5', $text);
}
```

As the `encrypt` method is used by the `exchangeArray` method of the same class, we also have to modify this last one, changing the `$this` keyword to `self`:

```
public function exchangeArray($data)
{
    $this->ID = ($data['ID'] ?? 0);
    $this->name = ($data['name'] ?? '');
    $this->nickname = ($data['nickname'] ?? '');
    $this->password = (isset($data['password']) ?
        self::encrypt($data['password']) : '');
}
```

Well, now we can use the `encrypt` method in the `IndexController->loginAction` method. Next, we have to replace the following line:

```
$this->adapter->setCredential($password);
```

It needs to be replaced with this:

```
$this->adapter->setCredential(Employee::encrypt($password));
```

Try to log in again. Again, you are disappointed; you see the **Invalid data** message again. What is wrong? Well, if you look at the `Employee::encrypt` method carefully, you will notice that it generates different hashes for the same password because there is a random component. Random components are very good to suggest passwords to users, but they are not appropriate for comparisons when the encrypted data needs to be the same. Thus, when you try to log in, the generated hash is different from the database stored hash.

Because of this, we have to modify the `Employee::encrypt` method for it to generate the same output for the same input. Let us modify the `Employee::encrypt` method with the following content:

```
public static function encrypt(String $text)
{
    $text = strrev(hash('sha256', $text));
    return hash('md5', $text);
}
```

To regenerate the password hash, you need to edit the employee Steven Grant at `http://localhost:8000/inventory/employee`. Now, type the same password, `moonknight`. In sequence, go back to the login page at `http://localhost:8000/inventory`, fill in the login form, and click on the **login** button. This time, the result will be the message shown in *Figure 9.10*:

Figure 9.10 – The message for a successful employee login

After you click on the **OK** button, you will see an error page, as shown in *Figure 9.11*:

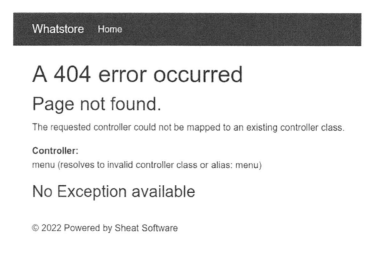

Figure 9.11 – The error page showed after login

This was expected because we didn't create the `MenuController` class and the view file for the `index` action. To solve this, we have to follow these steps:

1. Create a `MenuController` class in the `module/Inventory/src/Controller` folder:

MenuController.php

```php
<?php
namespace Inventory\Controller;
use Laminas\Mvc\Controller\AbstractActionController;
use Laminas\View\Model\ViewModel;

class MenuController extends AbstractActionController
{
    public function indexAction()
    {
        return new ViewModel();
    }
}
```

2. Add the following entry to the `controllers => aliases` key in the `modules.config.php` file:

```php
'menu' => Controller\MenuController::class
```

3. Add the following entry to the `controllers => factories` key in the `modules.config.php` file:

```php
Controller\MenuController::class => InvokableFactory
    ::class
```

4. Change the value of the `router => routes => inventory => options => defaults => controller` key with index like so:

```php
'controller' => 'index'
```

5. Create an `index.phtml` file in the `module/Inventory/view/inventory/menu` folder (you need to create the menu folder):

```html
<h1>Inventory Module</h1>

<a href="<?=$this->url('inventory', ['action' =>
    'logout']) ?>">Logout</a>
```

6.   After these changes, reload the page at `http://localhost:8000/inventory/menu`, and you will see the page shown in *Figure 9.12*:

Figure 9.12 – The menu page for Inventory Module

Now, if you click on the **Logout** option, you will go back to the login form page. Great! It seems that it is working. However, if you try to access any other page of the `Inventory` module now, such as `http://localhost:8000/inventory/menu`, you will notice that you don't need to be authenticated to access any page. Then, the login page is useless. This is because we haven't verified the user identity yet. But don't worry! We will do that in the next section.

Verifying user identity based on events

To verify the identity of a user and know whether they are logged in, we will use a `listener` class. As we have seen in *Chapter 7, Request Control and Data View*, Laminas uses event-driven programming to process the steps of a request life cycle. There are some defined events, and we can create listeners to act when these events are triggered.

As we don't need to store data for this verification, we will use a `static` method (a class method). Therefore, we don't need to instantiate a class. We will create a new folder named `Listener` in `modules/Inventory/src`. Inside this new folder, we will create a class named `InventoryAuthenticationListener`. This class will have a method named `verifyIdentity`, as shown in the following source code in sequence:

InventoryAuthenticationListener.php

```php
<?php
namespace Inventory\Listener;

use Laminas\Mvc\MvcEvent;
use Laminas\Authentication\AuthenticationService;
```

```php
use Laminas\Http\PhpEnvironment\Request;

class InventoryAuthenticationListener
{
    public static function verifyIdentity(MvcEvent $event)
    {
        $routeName = $event->getRouteMatch()->
            getMatchedRouteName();
        if (!$routeName == 'inventory'){
            return;
        }
        $authenticationService = new AuthenticationService
            ();
        if (!$authenticationService->hasIdentity()){
            $params = $event->getRouteMatch()->getParams();
            $controller = $params['controller'];
            $action = $params['action'];
            if ($controller == 'index' && ($action ==
                'index' || $action == 'login')){
                return;
            }
            $event->getRouteMatch()->setParam('controller',
                'index');
            $event->getRouteMatch()->setParam('action',
                'index');
        }
    }
}
```

Let us see what `InventoryAuthenticationListener::verifyIdentity` does:

- The `verifyIdentity` method receives a `MvcEvent` object. The `MvcEvent` class is a specialization of the `Laminas\EventManager\Event` class for Laminas MVC implementation. It means that `MvcEvent` class has the generic implementation of the `Event` class for any kind of event and specific implementation for MVC events. An `MvcEvent` object has access to the `Application` instance and every object configured in the `modules.config.php` file.

- We use the `getRouteMatch` method of `MvcEvent` to recover a `RouteMatch` object. This object stores data about the matched route. The `RouteMatch->getMatchedRouteName` method returns the name of the route that matches the requested URL.

- We use the route name to know whether the requested module is `Inventory`. As `InventoryAuthenticationListener` is only for the `Inventory` module, we abort the `verifyIdentity` method if the request is from another module.

- If the requested module is `Inventory`, we instantiate the `AuthenticationService` class and invoke the `hasIdentity` method. This method returns `true` only if a user has been authenticated.

- If a user has not been authenticated, we recover the controller and action name from the matched route through the `getParams` method of the `RouteMatch` instance.

- The employee – that is, the inventory user – can access the `IndexController->indexAction` and `IndexController->loginAction` methods without authentication, and then these two methods are ignored by the authentication listener class.

- For any other controller and action of the `Inventory` module, if the user is not authenticated, we redirect them to the login page.

Note that we are intercepting the application request before a controller is instantiated. We don't need to make an HTTP request to redirect the user to the login page. The target change is made along the same HTTP request.

However, who will invoke `InventoryAuthenticationListener`? A listener class is invoked by an event. We associate listeners and events with the `Laminas\EventManager\EventManager` class. The `Laminas\Mvc\Application` object, created in the `index.php` file, has an instance of `Laminas\EventManager\EventManager`. We will use this instance to add `InventoryAuthenticationListener` as a listener for the `EVENT_ROUTE` event. The `EVENT_ROUTE` event is dispatched after the route is matched, so we have all the necessary data about the desired resources of the user and the initial target of the request.

The last line of the `index.php` file in the `public` folder is this:

```
Application::init($appConfig)->run();
```

Let us modify this preceding line, dividing the method invocations with these two steps:

```
$app = Application::init($appConfig);
$app->run();
```

We are doing the same thing in these lines that we did in the preceding line but in different lines. The instance of `Laminas\Mvc\Application` is stored now in the `$app` variable. Let us introduce a new line between them:

```
$app = Application::init($appConfig);
$app->getEventManager()->attach(MvcEvent::EVENT_ROUTE,
[InventoryAuthenticationListener::class,'verifyIdentity']);
$app->run();
```

What are we doing here? The `attach` method of `EventManager` associates a listener to an event. In other words, when the specified events are dispatched, the listener is notified. A listener can be a static method, an object method, or a function. For static methods, we need to pass an array with two elements – the class name and the method name. For the object method, we need to pass the instance as the first element of the array. For a function, only the function name is needed as the second argument of the `attach` method.

> **Remember, remember the namespaces!**
>
> Remember that `InventoryAuthenticationListener` is defined for the `Inventory\Listener` namespace, so, your `index.php` file must have use `Inventory\Listener\InventoryAuthenticationListener`; at the top. The lack of namespaces is a common source of errors.

Let us modify the menu view to show the user identity. Change the content of the `MenuController->indexAction` method of the `Inventory` module with the following content:

```
    public function indexAction()
    {
        $auth = new AuthenticationService();
        return new ViewModel(['user' => $auth->
            getIdentity()]);
    }
```

The `getIdentity` method of the `AuthenticationService` class returns a matched identity in the authentication. In sequence, change the content of the `index.phtml` file in the `module/Inventory/view/inventory/menu` folder with the following content:

```
<h1>Inventory Module</h1>

<p>Hi, <?=$this->user?></p>
<a href="<?=$this->url('inventory',['action' =>
    'logout']) ?>">Logout</a>
```

Now, you can log in again. This time, you will see the user's name on the menu page, as shown in *Figure 9.13*:

Figure 9.13 – The menu page with the identity of the user

Click on the **Logout** option and try to access any `Inventory` module page, such as employee registration at `http://localhost:8000/inventory/employee` or product registration at `http://localhost:8000/inventory/product`. You will see that you will always be redirected to the login page. It is because you are not authenticated. Now, access to the `Inventory` module is only for employees.

It seems that authentication has been implemented successfully. However, then you run some tests and see errors like so:

```
1) InventoryTest\Controller\
IndexControllerTest::testIndexActionCanBeAccessed
Failed asserting controller name "inventory\controller\
indexcontroller", actual controller name is "index"

/opt/lampp/htdocs/whatstore/vendor/laminas/laminas-test/src/
PHPUnit/Controller/AbstractControllerTestCase.php:611
/opt/lampp/htdocs/whatstore/module/Inventory/test/Controller/
IndexControllerTest.php:34

2) InventoryTest\Controller\
IndexControllerTest::testLoginActionCanBeAccessed
Failed asserting response code "302", actual status code is
"200"
```

```
/opt/lampp/htdocs/whatstore/vendor/laminas/laminas-test/src/
PHPUnit/Controller/AbstractControllerTestCase.php:434
/opt/lampp/htdocs/whatstore/module/Inventory/test/Controller/
IndexControllerTest.php:46


3) InventoryTest\Controller\
IndexControllerTest::testLogoutActionCanBeAccessed
Failed asserting response code "200", actual status code is
"302"


/opt/lampp/htdocs/whatstore/vendor/laminas/laminas-test/src/
PHPUnit/Controller/AbstractControllerTestCase.php:434
/opt/lampp/htdocs/whatstore/module/Inventory/test/Controller/
IndexControllerTest.php:56
```

Let us fix them, one by one:

- The error in the `IndexControllerTest::testIndexActionCanBeAccessed` method occurs because we have changed the value for the default controller of the inventory route in `module.config.php`.

  You must replace the following line in `IndexControllerTest::testIndex ActionCanBeAccessed`:

  ```
  $this->assertControllerName(IndexController::class);
  ```

  Replace it with this:

  ```
  $this->assertControllerName('index');
  ```

  In fact, the correction for this error was indicated in the *as specified in router's controller name alias* comment that appears after the `assertControllerName` method is called. We have copied this line and this comment in the *Implementing CRUD with a controller and view pages* section of *Chapter 7, Request Control and Data View*.

- The error in the `InventoryTest\Controller\IndexControllerTest::testLogin ActionCanBeAccessed` method occurs because `IndexController->loginAction` returns an HTTP 200 status code. It is JavaScript that redirects us to the menu page, but we are not testing JavaScript here.

  Change the `assertResponseStatusCode` argument to 200, as per the following code line:

  ```
  $this->assertResponseStatusCode(200);
  ```

- The error in the `InventoryTest\Controller\IndexControllerTest::testLogout`
  `ActionCanBeAccessed` method occurs because `IndexController->logoutAction`
  redirects us to the login page, returning an HTTP 302 status code.

  To resolve this, change the `assertResponseStatusCode` argument to 302, as per the
  following code line:

  ```
  $this->assertResponseStatusCode(302);
  ```

It's ready. You can now run the tests successfully. Congratulations! You have now implemented authentication in the `Inventory` module! You may have thought it was more difficult, but now you can see that it is not complicated.

Well, we could stop here, but there is an appropriate change to make in the `index.php` file before ending the discussion about authentication. In the next section, we will learn how to transfer the responsibility for listening to the `EVENT_ROUTE` event from `index.php` to the `Inventory` module.

Delegating the authentication listener to a module

In the previous section, you learned that it is possible to call other methods from the `Application` instance besides the `run` method. Before inverting the application control from your code to Laminas, it is possible to modify the behavior through the `Application` class.

However, the `index.php` file, where `Application` is instantiated, is not the appropriate place to modify every application behavior. In an object-oriented application, we need to delegate each action to an appropriated class. When we attach an event listener in `index.php`, we are delegating the responsibility to manage events to this file. The question is whether `index.php` should be responsible for this.

Think about it – who is responsible for ensuring that the `Inventory` module has authentication control? Maybe this question could be not the best. Let us ask another – is authentication control always required for the `Inventory` module? Yes! Access will always be restricted. So, it seems appropriate that the `Inventory` module takes care of all its own authentication control. If this module is reused in another project, listening for authentication control will already be embedded.

Right, how can we transfer the responsibility to attach the event listener for authentication from `index.php` to the `Inventory` module? The first step is rolling back the change in `index.php`. This means that the following lines should be replaced:

```
$app = Application::init($appConfig);
$app->getEventManager()->attach(MvcEvent::EVENT_ROUTE,
[InventoryAuthenticationListener::class, 'verifyIdentity']);
$app->run();
```

They should be replaced by the following line:

```
$app = Application::init($appConfig)->run();
```

The next step is to add a new method in the Module class of the Inventory module. It is the onBootstrap method. This method is called when an EVENT_BOOTSTRAP event is triggered. It is a logical choice because an EVENT_ROUTE event is the next in the sequence:

```
/**
 * @param   \Laminas\Mvc\MvcEvent $e The MvcEvent
     instance
 * @return void
 */
public function onBootstrap($e)
{
    $application = $e->getApplication();
    $application->getEventManager()->attach
        (MvcEvent::EVENT_ROUTE,
 [InventoryAuthenticationListener::class, 'verifyIdentity']);
}
```

Observe that the onBootstrap method receives a MvcEvent object. This object has access to the Application instance – the same instance that was created in index.php. The second line of the onBootstrap method does exactly the same operation that was removed from index.php. This way, the transference of responsibilities is done. The Inventory module is responsible for its own authentication process. It does not depend on another component to ensure authentication.

We have changed the authentication implementation but not its behavior. You can now run the tests again to check whether everything is working. Unfortunately, when you run the tests, some failures are presented by PHPUnit. You will see several error messages like this:

```
Failed asserting controller name "discount", actual controller
name is "index"
```

In fact, you will notice that all error messages have the same pattern – the matched controller name is IndexController. Why?

The reason is that now InventoryAuthenticationListener is attached before the tests run. The tests don't use index.php. So, when the attachment was done by index.php, there was no authentication checking because index.php was not called by the PHPUnit test case. Now, as the listener attachment is done by the Inventory module, authentication checking is done for each request.

Does it mean we need to authenticate beforehand to test? Not really. We want to test units, not complete processes. It is enough, for now, to disable the authentication checking for testing context. This can be done by adding a conditional instruction to the `Module::onBootstrap` method in the `Inventory` module. We verify whether the `PHPUNIT_COMPOSER_INSTALL` constant was defined. This constant is defined by PHPUnit, installed as a Composer dependency, which it is in our case.

```
Public function onBootstrap($e)
{
    if (defined('PHPUNIT_COMPOSER_INSTALL')) return;
    $application = $e->getApplication();
    $application->getEventManager()->attach(MvcEvent::
        EVENT_ROUTE,
[InventoryAuthenticationListener::class, 'verifyIdentity']);
}
```

Finally, you can run the tests successfully!

Summary

In this chapter, we have learned how to implement a checking process to authenticate users with event-oriented programming.

First, we created a login page for the `Inventory` module, handled by the `IndexController` class. In sequence, we installed the `laminas-authenticator` component and used it to authenticate the `Inventory` module users – the employees. Then, we learned how to implement the employee registration pages by remembering what we learned in *Chapter 7, Request Control and Data View*, and *Chapter 8, Creating Forms and Implementing Filters and Validators*. Finally, we learned how to verify authenticated users with `laminas-eventmanager`.

In the next chapter, we will learn how to implement authorization for users using event-driven programming.

10

Event-Driven Authorization

This chapter covers one of the essential topics when developing secure applications: authorization.

Authorization is checking the allowed actions for an identified person. An example of authorization in the physical world is the control over the floors and rooms that a person can access after getting into a building. It can be carried out by a person, keys or cards, or even lifts that open to authorized floors only. In short, authorization is a limitation of actions that somebody can do with a set of resources.

In this chapter, you will learn how to implement authorization control with event-oriented programming. First, we will create registrations for roles and resources, the main concepts of the **Role-Based Access Control** (**RBAC**) approach. We will associate resources with roles and roles with users for, in sequence, implementing an authorization control for employees in the `Inventory` module. Finally, we will create a listener to verify user permissions for each request.

In this chapter, we'll be covering the following topics:

- Creating roles registration
- Creating resources registration
- Associating resources with roles
- Associating roles with users
- Implementing access control
- Verifying users' permissions
- Modifying the test bootstrap
- Creating an `identity` manager

By the end of this chapter, you will be able to control which resources each user can handle in an application.

Let us start by creating the registrations for roles and resources.

Technical requirements

All the code related to this chapter can be found at `https://github.com/PacktPublishing/PHP-Web-Development-with-Laminas/tree/main/chapter10/whatstore`.

> **Important note: Possible session crash**
>
> In this chapter, we will make changes in the data structure stored by application in the session. If you didn't log out from the application of *Chapter 9, Event-Driven Authentication* and try to log in the application of this chapter, you will be surprised with a session error. If this happens, clear the session before proceeding. If you don't want to clear the session through browser, you can use a script for this available at `https://github.com/PacktPublishing/PHP-Web-Development-with-Laminas/tree/main/util`.

Creating roles registration

Employees are the users of the `Inventory` module. Employees play different roles in a company over time. An employee can do tasks of a fellow employee who is on vacation. An employee may also take the place of their boss when the boss is on vacation. At different times, employees have different permissions according to the tasks they need to do. So, we mustn't strongly couple employees and permissions because permissions are more connected to roles.

In fact, imagine you have already implemented CRUD for roles, as we did in *Chapter 7, Request Control and Data View*, with the products. Also, imagine that you have implemented this CRUD using `laminas-form`, `laminas-filter`, and `laminas-validation`, as we did in *Chapter 8, Creating Forms and Implementing Filters and Validators*, for product discounts. It is not difficult to imagine because the source code is actually available in the repository mentioned in the *Technical requirements* section. As we are using patterns, the structure of CRUD implementations is the same. Yes, we're getting déjà vu.

So, we are assuming that there are the following classes in the `module/Inventory/src/Model` folder:

- `Role`
- `RoleTable`
- `RoleTableFactory`

It is important to remember that the factories of each table gateway class must be configured by the `service_manager` key in the `modules.config.php` file.

We are also assuming that there are the following classes in the `module/Inventory/src/Controller` folder:

- `RoleAPIController`

- `RoleAPIControllerFactory`
- `RoleController`
- `RoleControllerFactory`

You can observe that we have two pairs of classes: an API controller and its factory, and a web interface controller and its factory. It is exactly the same structure that we used for product controller layer implementation. These controller classes are tested by the following classes in the `module/Inventory/test/Controller` folder:

- `RoleAPIControllerTest`
- `RoleControllerTest`

It is important to remember that the aliases and factories of each controller class must be configured by the `controllers` key in the `modules.config.php` file.

We also have a view layer implementation for roles that is very similar to what we had for products. So, we are assuming that the following view scripts are in the `module/Inventory/view/inventory/employee` folder:

- `edit.js`
- `edit.phtml`
- `index.js`
- `index.phtml`

Here, we have two pairs of files that generate two web pages: the inserting/updating page and the listing page. The `.phtml` files have a combination of HTML and PHP code and the `.js` files have exclusively JavaScript code.

Right, we now have role registration. This registration is simpler than employee registration because there are only two attributes, code and name. In the next section, we will create resources registration.

Creating resources registration

When we are dealing with permissions, we are talking about someone that can do some action with something. In the model that we are working with, who does the action is the role. What is impacted by the action of a role is a resource.

Roles are perhaps easier to understand because you can quickly establish a connection between users and roles. You can imagine real people when we are talking about users and roles.

But what is a resource? It seems that we are talking about economics... No, for an access control system, a resource is anything that a role can handle. Resources can be objects, URLs, visual components, and so on. It depends on the system. Or better, it depends on the system requirements. Resources are a part of system requirements. There must be a system in place to determine who can do what. It is part of the business. So, the customer of the system needs to specify what they expect in terms of access control.

Do you remember Mr. Heythere, the CEO of Whatstore? Yes, someone has hired Sheat Software to build an awesome e-commerce website. Well, Sheat Software architect Mr. Worried talked to Mr. Heythere about employee permissions. Mr. Heythere didn't have a definite idea about how to control access and asked for suggestions from Mr. Worried. Mr. Worried explained a little about the structure of Laminas using the medieval castle metaphor. Mr. Heythere liked the concept of controllers and actions. He found them interesting. Mr. Heythere decided that resources for the Whatstore system could each be a combination of a controller and action for the `Inventory` module.

Imagine that, after defining the resource definition, you implement CRUD for resources, in a similar way that we did for roles in the previous section. You should understand the MVC implementation for registrations. So, we won't repeat that. Let us focus on how to fill the `resources` table.

Filling the resources table

In fact, to give permissions, we need at least one registered user (employee), associated with one role and registered resources. We can provide these conditions using a script to prepare the database with some essential records. We will create a script named `loaddatabase.php` in the `bin` folder of the `whatstore` project. As we saw in *Chapter 5*, *Creating the Virtual Store Project*, the `bin` folder stores automation scripts.

The `load_database.php` script file begins with the dependencies declaration:

```
<?php
use Laminas\Db\Adapter\Adapter;
use Laminas\Stdlib\ArrayUtils;
use Inventory\Model\Employee;
use Laminas\Db\Adapter\AdapterInterface;


require __DIR__ . '/../vendor/autoload.php';
```

In the preceding code block, we are using the `Adapter` class and the `AdapterInterface` interface from `laminas-db` to create an object that abstracts the database connection. The `ArrayUtils` class serves to merge configuration arrays properly. We need to have an explicit reference to the `Employee` class because of password encryption. For all these classes and interfaces to be found, we have to import the Composer autoload file (`autoload.php`). In sequence, we will declare a function (`getConfig`) to get the application configuration, like so:

```
function getConfig(): array
{
```

```
    $config = require __DIR__ . '/../config/autoload/
        global.php';

    if (file_exists(__DIR__ . '/../config/autoload/
        local.php')){
        $override = require __DIR__ . '/../config/autoload/
            local.php';
        $config = ArrayUtils::merge($config, $override);
    }
    return $config;
}
```

This getConfig function has the same code lines used in the index.php file in the public folder. This function partially emulates the application initialization. In sequence, we will declare another function (recreateDatabase) to drop the database and recreate it with some essential data:

```
function recreateDatabase(AdapterInterface $adapter): void
{
    echo "Dropping database...\n";
    $adapter->query('DROP DATABASE IF EXISTS whatstore',
        Adapter::QUERY_MODE_EXECUTE);
    $sqlscript = file_get_contents(__DIR__ .
        '/whatstore.sql');
    echo "Creating database and tables...\n";
    $adapter->query($sqlscript, Adapter::
        QUERY_MODE_EXECUTE);
    echo "Whatstore database has been created.\n";
}
```

The recreateDatabase function uses an AdapterInterface implementation to drop the database and recreate it in sequence using a SQL script (which can be found in the GitHub repository). In sequence, we declare a function (addUserAndRole) to create and record in the employee table and another one in the roles table:

```
function addUserAndRole($adapter): void
{
    $password = Employee::encrypt('moonknight');
    $sqlscript = <<<SQL
INSERT INTO `employees`(`ID`, `name`, `nickname`,
```

```
    `password`) VALUES (NULL,'STEVEN GRANT','MARK SPEKTOR'
        ,'$password');
SQL;
    $adapter->query($sqlscript, Adapter::
        QUERY_MODE_EXECUTE);
    $sqlscript = <<<SQL
INSERT INTO `roles`(`code`, `name`) VALUES (NULL,'ADMIN');
SQL;
    $adapter->query($sqlscript, Adapter::
        QUERY_MODE_EXECUTE);
    $sqlscript = <<<SQL
    INSERT INTO `employee_roles`(`code`, `code_role`,
    `ID_employee`) VALUES (NULL,(SELECT code FROM `roles`
     WHERE name = 'ADMIN'),(SELECT ID FROM `employees`
     WHERE name = 'STEVEN GRANT'));
SQL;
    $adapter->query($sqlscript, Adapter::
        QUERY_MODE_EXECUTE);

    echo "A user and a role were created.\n";
}
```

The addUserAndRole function uses an AdapterInterface implementation to insert an employee, insert a role, and then associate this employee and this role in the employee_roles table. In sequence, we will declare a function (addResource) to extract data about Inventory controllers and their methods to register resources:

```
function addResources($adapter): void
{
    $dir = scandir(realpath(__DIR__ . '/../module/Inventory
        /src/Controller'));

    $controllers = [];

    foreach($dir as $file){
        if (strpos($file,'Controller.php') !== false){
            $controllers[] = 'Inventory\Controller\\' . str_
replace('.php','',$file);
```

```
        }
    }

    $resources = getResources($controllers);

    $sqlscript = <<<SQL
INSERT INTO `resources`(`code`, `name`, `method`) VALUES
    $resources;
SQL;

    $adapter->query($sqlscript, Adapter::
        QUERY_MODE_EXECUTE);

    echo "Resources were created.\n";
}
```

What does the addResources function do? Let us understand:

- First, this function scans the controller directory and creates a list with the complete qualified name of all Inventory controllers.

- Then, this function calls the getResources function to get a list of resources.

The getResources function receives an array, which is the list of Inventory controllers. The code of this function is as follows:

```
function getResources(array $controllers): String
{
    $actionMethods = ['getMethodFromAction',
        'notFoundAction'];
    $httpMethods = ['create','update','get','delete'];
    $restMethods = [
        'create' => 'POST',
        'update' => 'PUT',
        'get' => 'GET',
        'delete' => 'DELETE'
    ];

    $resources = '';
```

```
foreach ($controllers as $controller){
    $methods = get_class_methods($controller);
    foreach($methods as $method){
        $alias = str_replace('Inventory\Controller\\',
            '',$controller);
        $alias = strtolower(str_replace('Controller',
            '',$alias));
        if (strpos($method,'Action') !== false &&
            !in_array($method, $actionMethods)){
            $method = str_replace('Action', '',
                $method);
            $resources .= "(NULL,'$alias.$method',
                'GET'),";
            $resources .= "(NULL,'$alias.$method',
                'POST'),";
        }
        if (in_array($method, $httpMethods)){
            $resources .= "(NULL,'$alias.{$method}',
            '{$restMethods[$method]}'),";
        }
    }
}
```

Observe that the getResources function generates a list with the concatenation of the controller alias and method name. The selected methods are the ones that have the Action suffix (except for some methods inherited from a superclass) or have a specific name related to the REST architecture.

The resource list is concatenated to a SQL INSERT statement, which is executed in sequence. At the end of the addResources function, the table resources are populated with Inventory resources whose access must be controlled.

The grantPermissionsToRole function uses an AdapterInterface implementation to grant the admin role permission to access all the resources registered by the addResources function. In the following code block, we can see the source code of the grantPermissionsToRole function:

```
function grantPermissionstoRole(AdapterInterface $adapter):
void
{
    $statement = $adapter->query('SELECT code FROM
`resources`');
```

```
        $results = $statement->execute();

        foreach($results as $row){
            $sqlscript = <<<SQL
        INSERT INTO `resources_role`(`code`, `code_role`,
        `code_resource`) VALUES (NULL,(SELECT code FROM `roles`
         WHERE name = 'ADMIN'),{$row['code']});
    SQL;
            $adapter->query($sqlscript, Adapter::
                QUERY_MODE_EXECUTE);
        }

        echo "Permissions were granted.\n";
    }
```

These functions are orchestrated by the following script:

```
// main program

$config = getConfig();
$db = $config['db'];
unset($db['database']); // remove database key because it
causes error in connection when database does not exist
$adapter = new Adapter($db);

try {
    recreateDatabase($adapter);
    $db = $config['db'];
    $adapter = new Adapter($db);
    addUserAndRole($adapter);
    addResources($adapter);
    grantPermissionsToRole($adapter);
} catch (Exception $e) {
    echo 'Fail to recreate the database.' .
        $e->getMessage() . "\n";
}
```

Why did we create this script? Because we need to have a user who can be authenticated and authorized. Now, how can we navigate through the `Inventory` module without a valid user or a user with permissions? You can run the following script inside the `bin` directory:

```
php load_database.php
```

At the end of the execution, you will have a user (an employee), a role, resources, and permissions for the user to access all these resources. Now, you can navigate through the `Inventory` module after the authorization control is enabled later in this chapter. But before we do this, we need to implement registrations to associate roles with resources and users with roles. The `load_database.php` script file made these registrations and associations for only one user, Steven Grant, who is, as a result, our superuser or administrator user. But the administrator needs an interface to associate a role with a user and grant permissions to users. Let us see how to implement the association between roles and resources in sequence in the next section.

Associating resources with roles

We won't repeat the details about the registration that we have already done. The code for the implementation of role resources is available in the source code folder of this chapter, mentioned in the *Technical requirements* section. Here, we will focus only on the differences related to the association between resources and roles. The registration for role resources is based on the `resources_role` table.

We have a `RoleResourceController` class for handling two pages: one for listing and deleting the pairs of resources and roles (`index.phtml`) and another one for inserting pairs of resources and roles (`edit.phtml`). To carry out these actions, the `RoleResourceController` constructor needs to receive three table gateway classes, as in the following snippet:

```
public function __construct(RoleResourceTable
$roleResourceTable, ResourceTable $resourceTable,
RoleTable $roleTable)
{
    $this->roleResourceTable = $roleResourceTable;
    $this->resourceTable = $resourceTable;
    $this->roleTable = $roleTable;
}
```

The `indexAction` method of the `RoleResourceController` class is similar to the previous controllers. This method provides an HTML page with permissions of roles to resources, as shown in *Figure 10.1*. This page is the output of a request to `http://localhost:8000/inventory/roleresource`:

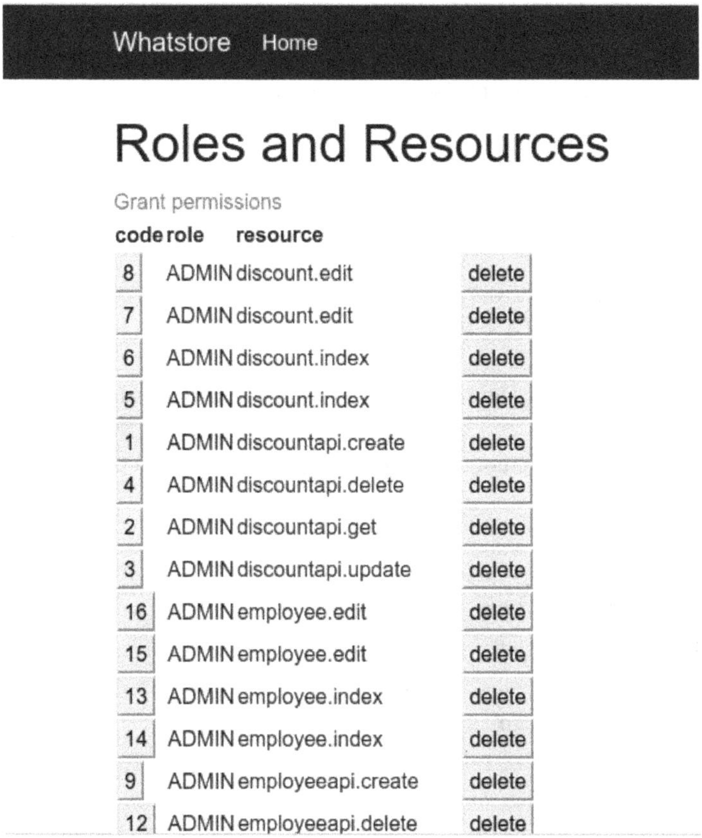

Figure 10.1: Roles and their resources

The page shown in *Figure 10.1* allows you to remove permissions for roles. As you can see, the list is sorted by role name and not by the permission code. This list is generated by the `RoleResourceTable->getAll` method. This method builds and executes a SQL `SELECT` statement that joins the `resources_role`, `resources`, and `roles` tables. The `getAll` method is listed in the sequence. In fact, you already saw the use of the `INNER JOIN` clause with the `Select` class in the *Testing the interaction between the web interface and API* section of *Chapter 7, Request Control and Data View*. The new addition to the following source code is the use of an `order` method, which generates the `ORDER BY` class of SQL:

```
public function getAll($where = null): iterable
{
    $select = new Select($this->tableGateway->
        getTable());
    $select->join('resources','resources_role.code_
```

```
            resource=resources.code',['resource' => 'name'])
        ->join('roles','resources_role.code_role=
            roles.code',['role' => 'name'])
        ->order(['roles.name','resources.name']);
        if (!is_null($where)){
            $select->where($where);
        }
        return $this->tableGateway->selectWith($select);
    }
```

You can see that the page shown in *Figure 10.1* has a **Grant permissions** hyperlink. This hyperlink invokes the RoleResourceController->editAction method. This method uses two table gateway objects to populate two different drop-down lists, as we can see in the following source code:

```
public function editAction()
{
    $form = new RoleResourceForm();
    $resources = [];
    $rows = $this->resourceTable->getAll();
    foreach($rows as $row){
        $resources[$row->code] = $row->name;
    }
    $roles = [];
    $rows = $this->roleTable->getAll();
    foreach($rows as $row){
        $roles[$row->code] = $row->name;
    }
    $form->get('code_resource')->setValueOptions
        ($resources);
    $form->get('code_role')->setValueOptions($roles);
    return new ViewModel([
        'form' => $form
    ]);
}
```

This method provides an HTML page with the form shown in *Figure 10.2*:

Figure 10.2: Form to grant permissions to roles

The form shown in *Figure 10.2* allows you to select a role and grant it permission to access a resource. If you click on the **save** button, the association will be persisted (if it does not exist yet). The persistence is made by the `RoleResourceTable->save` method, as in the following source code:

```php
public function save(AbstractModel $model): bool
{
    $set = $model->toArray();
    $resourcesRole = $this->getByFields([
        'code_resource' => $model->resource->code,
        'code_role'=> $model->role->code,
    ]);
    try {
        if (empty($resourcesRole->code)) {
            $this->tableGateway->insert($set);
        }
    } catch (\Exception $e) {
        error_log($e->getMessage());
        return false;
    }
    return true;
}
```

Notice that this save method calls a getByFields method. The getByFields method was added to the AbstractTable class in the Generic module to allow queries with several filter conditions. The getByFields method receives an array as a parameter, as we can see in the following source code:

```
public function getByFields(array $fields): Abstract
    Model
{
    $where = [];
    foreach($fields as $field => $value){
        $where[$field] = $value;
    }
    $rowSet = $this->getAll($where);
    if ($rowSet->count() == 0) {
        $modelName = $this->modelName;
        return new $modelName();
    }
    return $rowSet->current();
}
```

Now, you are able to associate resources with a role and so allow a role to have access to these resources.

In the next section, we will see how to implement the association between roles and users.

Associating roles with users

I won't repeat the details about registration that we have already covered. The code for the implementation of employee roles is available in the source code folder of this chapter, mentioned in the *Technical requirements* section. Here, we will focus only on the differences related to the association between roles and users of the Inventory module, the employees. The registration for employee roles is based on the employee_roles table.

Well, now that we have implemented the last registration of the Inventory module, it is time to add hyperlinks to the menu page for our registrations. So, we have to modify the index.phtml file in the module/Inventory/view/inventory/menu folder, like so:

index.phtml

```
<h1>Inventory Module</h1>
<p>Hi, <?=$this->user?></p>
<ul>
<li><a href="<?=$this->url('inventory',[
    'controller' => 'product'
])?>">Product Registration</a></li>
<li><a href="<?=$this->url('inventory',[
    'controller' => 'discount'
])?>">Discount Registration</a></li>
<li><a href="<?=$this->url('inventory',[
    'controller' => 'employee'
])?>">Employee Registration</a></li>
<li><a href="<?=$this->url('inventory',[
    'controller' => 'role'
])?>">Role Registration</a></li>
<li><a href="<?=$this->url('inventory',[
    'controller' => 'resource'
])?>">Resource Registration</a></li>
<li><a href="<?=$this->url('inventory',[
    'controller' => 'roleresource'
])?>">Resources of a role</a></li>
<li><a href="<?=$this->url('inventory',[
    'controller' => 'employeerole'
])?>">Roles of an employee</a></li>
</ul>

<a href="<?=$this->url('inventory',['action' => 'logout'
    ])?>">Logout</a>
```

This `index.phtml` file will show the following page:

Figure 10.3: Menu page for the Inventory module

The **Resources of a role** hyperlink takes you to a listing page of the registration presented in the previous section. The **Roles of an employee** hyperlink will lead you to the following page:

Figure 10.4: Listing page for employee roles

The page shown in *Figure 10.4* allows you to remove roles for employees. This list is generated by the `EmployeeResourceTable->getAll` method. This method builds and executes a SQL `SELECT` statement that joins the `resources_role`, `resources`, and `roles` tables. The `getAll` method is listed in the sequence. It is another use of the INNER JOIN clause example, as was presented in the previous section.

In the following code block, we can see the source code of the EmployeeResourceTable->getAll method:

```
public function getAll($where = null): iterable
{
    $select = new Select($this->tableGateway->
        getTable());
    $select->join('roles','employee_roles.
        code_role=roles.code',['role' => 'name'])
        ->join('employees','employee_roles.ID_
employee=employees.ID'
    ,['employee' => 'name'])
        ->order(['employees.name','roles.name']);
    if (!is_null($where)){
        $select->where($where);
    }
    return $this->tableGateway->selectWith($select);
}
```

As you can see, the page shown in *Figure 10.4* has an **Add a role to an user** hyperlink. This hyperlink invokes the EmployeeRoleController->editAction method. This method uses two table gateway objects to populate two different drop-down lists, as we can see in the following source code:

```
public function editAction()
{
    $form = new EmployeeRoleForm();
    $employees = [];
    $rows = $this->employeeTable->getAll();
    foreach($rows as $row){
        $employees[$row->ID] = $row->name;
    }
    $roles = [];
    $rows = $this->roleTable->getAll();
    foreach($rows as $row){
        $roles[$row->code] = $row->name;
    }
```

```
$form->get('ID_employee')->setValueOptions
    ($employees);
$form->get('code_role')->setValueOptions($roles);
return new ViewModel([
    'form' => $form
]);
}
```

This method provides an HTML page with the form shown in *Figure 10.5*:

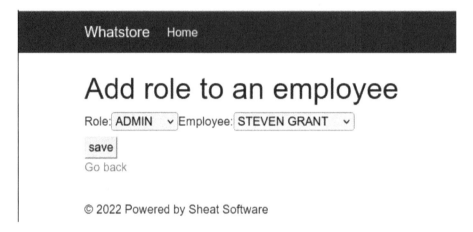

Figure 10.5: Form to associate roles with an employee

The form shown in *Figure 10.5* allows you to select a role and grant it permission to access a resource. If you click on the **save** button, the association will be persisted (if it does not exist yet). The persistence is made by the EmployeeRoleTable->save method, as in the following source code:

```
public function save(AbstractModel $model): bool
{
    $set = $model->toArray();
    $employeeRole = $this->getByFields([
        'code_role'=> $model->role->code,
        'ID_employee' => $model->employee->ID
    ]);
    try {
        if (empty($employeeRole->code)) {
            $this->tableGateway->insert($set);
        }
```

```
    } catch (\Exception $e) {
        error_log($e->getMessage());
        return false;
    }
    return true;
}
```

Now we know how to associate resources with roles and roles with employees. In the next section, we will see how to implement access control using these registrations.

Implementing access control

Laminas has two components for authorization: laminas-permissions-acl and laminas-permissions-rbac. The first component implements the **Access Control List** (**ACL**) approach, and the second component implements the RBAC approach.

We will use the laminas-permissions-rbac component to implement an authorization control in the whatstore project. Our system emphasizes roles in its data model and RBAC seems convenient. Just as we have added and installed the laminas-db component using Composer, we will do the same for the laminas-permissions-rbac component. Open the composer.json file in Eclipse and from the **Dependencies** tab, search for the laminas/laminas-permissions-rbac package and add it. Don't forget to save the file and click on the **Update dependencies** button. If you are unsure about how to install Composer dependencies with Eclipse, read the *Creating our first automated test* section in *Chapter 3, Using Laminas as a Library with Test-Driven Development*.

After installing laminas-permissions-rbac, we can use this component to modify the IndexController->loginAction method. This method will look like this:

```
public function loginAction()
{
    $nickname = $this->request->getPost('nickname');
    $password = $this->request->getPost('password');
    $this->adapter->setIdentity($nickname);
    $this->adapter->setCredential
        (Employee::encrypt($password));

    $logged = false;

    $auth = new AuthenticationService();
```

```
$auth->setAdapter($this->adapter);

$result = $auth->authenticate();

if ($result->isValid()) {
    error_log('user ' . $result->getIdentity() . '
        is logged');
    $this->loadAuthorizationData($result->
        getIdentity());
    $logged = true;
} else {
    foreach ($result->getMessages() as $message) {
        error_log($message);
    }
}

return new JsonModel([
    'logged' => $logged
]);
}
```

The only difference between the implementation presented here and in the previous chapter is that here, we call the loadAuthorizationData method after a successful authentication. The source code of this method is as follows:

```
private function loadAuthorizationData(string
    $identity): void
{

$role = new Role($identity);

$row = $this->adapter->getResultRowObject();
$resources = $this->resourceTable->
    getResourcesFromAnEmployee($row->ID);
foreach($resources as $resource){
    $role->addPermission($resource->name . ':' .
        $resource->method);
}
```

```
    $rbac = new Rbac();
    $rbac->addRole($role);

    $container = new Container();
    $container->rbac = $rbac;
}
```

The `loadAuthorizationData` method does the following:

- Creates an instance of the `Laminas\Permissions\Rbac\Role` class and uses the employee nickname as the role name. Pay attention! An employee can have several roles, but to check authorization, we will work with a virtual role that will unify all the real roles associated with the employee.

- Recovers the record of the authenticated employee with the `getResultRowObject` method of the `Laminas\Authentication\Adapter\DbTable\AbstractAdapter` class.

- Uses the ID of an authenticated employee to recover the resources associated with the roles of the employee through the `ResourceTable->getResourcesFromAnEmployee` method, whose source code is as follows:

```
    public function getResourcesFromAnEmployee
        ($employeeID): ResultSet
    {
        $subSelect = new Select('employee_roles');
        $subSelect->columns(['code_role'])
        ->where(['ID_employee' => $employeeID]);

        $where = new Where();
        $where->in('resources_role.code_role',
            $subSelect);

        $select = new Select($this->tableGateway
            ->getTable());
        $select->columns(['name' => new Expression
            ('DISTINCT name'),'method'])
        ->join('resources_role', 'resources.code =
            resources_role.code_resource', [])
        ->where($where);
```

```
        return $this->tableGateway->selectWith
            ($select);
    }
```

Notice that the ResourceTable->getResourcesFromAnEmployee method uses an instance of the Laminas\Db\Sql\Where class to define a filter with the SQL IN clause. Note also that this method uses the Laminas\Db\Sql\Expression class to add the DISTINCT keyword to the name field of the resources table. The use of DISTINCT is to avoid repeated resources when an employee has more than one role and two or more roles have permission to access the same resources.

- Creates an instance of the Laminas\Permissions\Rbac\Rbac class and associates the role object with this instance.

- Creates an instance of the Laminas\Session\Container class and stores the Rbac instance in the session.

So, we store the permissions of the authenticated employee in the session and can check them at any time. As we are storing the access data in the session when the user logs into the system, we need to remove this data when the user logs out of the system. We will use this in the IndexController->logoutAction method:

```
public function logoutAction()
{
    $auth = new AuthenticationService();
    $auth->clearIdentity();
    $session = new SessionManager();
    try {
        $session->destroy();
    } catch (\Exception $e) {
        error_log($e->getMessage());
    }

    return $this->redirect()->toRoute('inventory');
}
```

In the preceding code block, we used the Laminas\Session\SessionManager class to destroy the session. Notice that we did this inside a try/catch structure. Why? Because this method can be run by tests, which don't create sessions.

Now what? Will the permissions be checked going forward? Not yet. We need to create a listener class and associate it with the EVENT_ROUTE event, as we did for authentication in *Chapter 9, Event-Driven Authentication*. Let's go!

Verifying users' permissions

To get started, inside the module/Inventory/src/Listener folder, we will create a class named InventoryAuthorizationListener. This class will have a method named verifyPermission, as shown in the following code block:

```
public static function verifyPermission(MvcEvent $event)
{
    $routeName = $event->getRouteMatch()->
        getMatchedRouteName();
    if (!$routeName == 'inventory'){
        return;
    }
    $authenticationService = new AuthenticationService
        ();
    if ($authenticationService->hasIdentity()){
        $params = $event->getRouteMatch()->getParams();
        $controller = $params['controller'];
        $action = $params['action'];

        if ($controller == 'menu' || $action ==
            'logout'){
            return;
        }

        $method = $event->getRequest()->getMethod();
        $permission = $controller . '.' . $action . ':'
            . $method;

        $container = new Container();
        $rbac = $container->rbac;

        $role = $authenticationService->getIdentity();
        if (!$rbac->isGranted($role,$permission)){
```

```
                    $event->getRouteMatch()->setParam
                        ('controller', 'menu');
                    $event->getRouteMatch()->setParam('action',
                        'no-permission');
                    error_log("$role has not permission to
                        $permission");
                }
            }
        }
```

Let us see what InventoryAuthenticationListener::verifyIdentity does:

- The verifyPermission method receives an MvcEvent object. We talked about this class in *Chapter 9, Event-Driven Authentication*.

- We used the getRouteMatch method of MvcEvent to recover a RouteMatch object. We also talked about this class in *Chapter 9, Event-Driven Authentication*.

- We used the route name to find out whether the requested module is Inventory as we did for the authentication listener.

- If the requested module is Inventory, we instantiate the AuthenticationService class and invoke the hasIdentity method. You already know about this method from *Chapter 9, Event-Driven Authentication*.

- If a user has been authenticated, we recover the controller and action name from the matched route through the getParams method of the RouteMatch instance.

- If the employee, that is, the inventory user, can access the IndexController->menuAction and IndexController->logoutAction methods without authorization, then we don't do anything for these methods.

- For any other controller and action of the Inventory module, if employee roles do not have permission to access the resource, we redirect the user to the no-permission page of IndexController. The no-permission.phtml file that creates this page must look like this:

no-permission.phtml

```
<h1>Inventory Module</h1>

<p>Hi, <?=$this->user?>. You don't have permission to
    requested resource.</p>
```

```
<a href="<?=$this->url('inventory', ['controller' =>
    'menu'])?>">Menu</a>
```

To enable the `InventoryAuthorizationListener` class as a listener, we have to modify the `Module.php` file of the `Inventory` module. We have to modify the `Module->onBootstrap` method for it, and it looks like this:

```php
public function onBootstrap($e)
{
    if (defined('PHPUNIT_COMPOSER_INSTALL')) return;
    $application = $e->getApplication();
    $application->getEventManager()->attach(MvcEvent::
        EVENT_ROUTE, [InventoryAuthenticationListener::
            class,'verifyIdentity']);
    $application->getEventManager()->attach(MvcEvent::
        EVENT_ROUTE, [InventoryAuthorizationListener::
            class,'verifyPermission']);
}
```

After these changes, you will try to remove the permissions of your user (employee) and see that you are redirected to the no-permission page. It is important to understand permission changes will have an effect only after the user logs out and logs in again because the data in the session has not changed.

Wow, we are almost done. The authorization process has been implemented. In the next section, we will make an adjustment for tests.

Modifying the test bootstrap

Tests always have to run from the same state. But during development, we can change the data used by tests several times. We have created a script to load the necessary data to use the system with enabled authentication and authorization. Now, we will create a script to prepare the environment to run the test. This script will stay in the `bin` folder and will be named `testbootstrap.php`. The script will have the following code:

testbootstrap.php

```php
<?php
use Laminas\Db\Adapter\Adapter;
use Laminas\Stdlib\ArrayUtils;
```

```php
require __DIR__ . '/../vendor/autoload.php';

$config = require __DIR__ . '/../config/autoload/global.php';

if (file_exists(__DIR__ . '/../config/autoload/
    local.php')){
    $override = require __DIR__ . '/../config/autoload/
        local.php';
    $config = ArrayUtils::merge($config, $override);
}

$db = $config['db'];
unset($db['database']); // remove database key because it
causes error in connection when database does not exist
$adapter = new Adapter($db);

try {
    echo "Dropping database...\n";
    $adapter->query('DROP DATABASE IF EXISTS whatstore',
        Adapter::QUERY_MODE_EXECUTE);
    $sqlscript = file_get_contents(__DIR__ .
        '/whatstore.sql');
    echo "Creating database and tables...\n";
    $adapter->query($sqlscript, Adapter::
        QUERY_MODE_EXECUTE);
    echo "The whatstore database has been created.\n";
} catch (Exception $e) {
    echo 'Fail to recreate the database.' . $e->
        getMessage() . "\n";
}
```

This script drops the database and recreates it from the whatstore.sql SQL script, available in the bin folder. So, the tests run over the void tables and keep them void after they have been run. Remember that we always delete the records that we add to tests.

To use testbootstrap.php with our PHPUnit tests, change the phpunit tag of the phpunit.xml file and add the bootstrap attribute, which will look like this:

```xml
<phpunit bootstrap="bin/testbootstrap.php" colors="true">
```

From now on, when you run the vendor/bin/phpunit command, the testbootstrap.php script will be run before the tests. Finally, we introduced time travel in our project, because now we can go back to the past whenever we want to test our application.

We now have authentication and authorization features working in the Inventory module. However, the implementation of these security features is not very appropriated here because this leaves a business responsibility inside a controller (IndexController of the Inventory module). We need to move this responsibility to a model. We will do this in the next section.

Creating an identity manager

In this section, we will refactor the authentication and authorization implementation to decouple these responsibilities from the controller layer.

The first step is to create a model class named IdentityManager in the Model folder of the Inventory module. This class will have four private attributes:

```
class IdentityManager
{
    private ?AdapterInterface $adapter;
    private ?ResourceTable $resourceTable;
    private ?array $encryptionMethod;
    private ?AuthenticationService $auth;
```

Let's understand the preceding code:

- $adapter is the instance of the authentication adapter class.
- $resourceTable is the instance of the ResourceTable class, needed for the authorization process.
- $encryptionMethod is the method to be applied to the user credential.
- $auth is the AuthenticationService instance, needed for the authentication process. Every instance of this class provides the same data because identity is stored in the user session.

The last attribute can be assigned internally. The values for the first three attributes, on the other hand, need to be injected:

```
public function __construct(AdapterInterface $adapter,
ResourceTable $resourceTable, array $encryptionMethod)
{
    $this->adapter = $adapter;
    $this->resourceTable = $resourceTable;
```

```
        $this->encryptionMethod = $encryptionMethod;
        $this->auth = new AuthenticationService();
    }
```

The IdentityManager class will have a login method, which will encapsulate most of the code from the IndexController->loginAction method:

```
public function login(string $identity, string
    $credential): bool
{
    $this->adapter->setIdentity($identity);
    $credential = call_user_func($this->
        encryptionMethod, $credential);
    $this->adapter->setCredential($credential);

    $this->auth->setAdapter($this->adapter);

    $result = $this->auth->authenticate();

    if ($result->isValid()) {
        error_log('user ' . $result->getIdentity() . '
            is logged');
        $this->auth->getStorage()->write($this->
        adapter->getResultRowObject(null,'password'));
        $this->loadAuthorizationData();
        return true;
    } else {
        foreach ($result->getMessages() as $message) {
            error_log($message);
        }
    }
    return false;
}
```

Observe that there are some differences introduced into the authentication process encapsulated by IdentityManager->login. The first difference is the use of an injected method to encrypt the credential. This change makes this class more flexible because there is no coupling with the Employee class. It is easier to modify the encryption method without writing into the IdentityManager class.

The second difference is the following line:

```
$this->auth->getStorage()->write($this->adapter->getResultRowOb
ject(null,'password'));
```

This instruction writes the user object (without the password) to the authentication session. It is necessary because we intend to later revalidate the permission data, so we need to store the user's primary key and not only the credential. This change will require some adjustments where identity is used, but only a few.

Similar to the previous method, the IdentityManager class will have a logout method, which will encapsulate most of the code from IndexController->logoutAction:

```
public function logout(): void
{
    $this->auth->clearIdentity();
    $session = new SessionManager();
    try {
        $session->destroy();
    } catch (\Exception $e) {
        error_log($e->getMessage());
    }
}
```

The IdentityManager class will encapsulate the IndexController->loadAuthorizationData method too. You can see a little difference in the recovery of the employee's primary key employee:

```
public function loadAuthorizationData(): void
{
    $role = new Role($this->auth->getIdentity()->name);

    $row = $this->auth->getStorage()->read();
    $resources = $this->resourceTable->
        getResourcesFromAnEmployee($row->ID);
    foreach($resources as $resource){
        $role->addPermission($resource->name . ':' .
            $resource->method);
    }
    $rbac = new Rbac();
```

```
        $rbac->addRole($role);

        $container = new Container();
        if (isset($container->rbac)){
            unset($container->rbac);
        }
        $container->rbac = $rbac;
    }
```

As you'll remember, this method was private in the `IndexController` class, but in `IdentityManager`, it is public. This change is to protect against a kind of attack that can be carried out by unhappy employees.

Imagine the following scenario: on a cold day in winter, Mr. Heythere finds that one of his managers, Mr. Barnaby, has not been performing as required. Mr. Heythere informs Mr. Barnaby of what is expected of him, but Mr. Barnaby answers that he would prefer not to do so. So, Mr. Heythere decides to fire Mr. Barnaby and find a better manager (probably someone who reads books by Packt Publishing!). The issue is that Mr. Barnaby knows that he still has access to the system after being fired because he is so familiar with it.

The terror sequence is as follows: Mr. Heythere fires Mr. Barnaby. Due to the session configuration, Mr. Barnaby is still logged in to the Whatstore system with his manager permissions. Even if another manager changes his permissions, this will only affect his next access attempt. Now, Mr. Barnaby is very mad, because he and Mr. Heythere obviously did not agree on the terms of his dismissal, and Mr. Barnaby wants Mr. Heythere to pay. Since Mr. Barnaby can still access the system with his previous privilege level, he does so and damages some records.

How can we avoid this terror story? One possibility is updating the user permissions while they are logged in. So, we have changed the visibility of the `loadAuthorizationData` method, to use it beyond the login process.

To finish the `IdentityManager` implementation, we will add three methods:

```
    public function getIdentity(): string
    {
        return $this->auth->getIdentity()->name;
    }
    public function hasIdentity(): bool
    {
        return $this->auth->hasIdentity();
    }
    public function getRbac(): Rbac
```

```
    {
        $this->loadAuthorizationData();
        $container = new Container();
        return $container->rbac;
    }
```

The getIdentity and hasIdentity methods are wrappers for homonymous methods of the AuthenticationService class. The getRbac method provides an Rbac instance with current permissions.

Great. Our IdentityManager class is ready. But as it has dependencies, we need a factory for it. So, create the IdentityManagerFactory class in the Model folder of the Inventory module:

```
class IdentityManagerFactory implements FactoryInterface
{
    public function __invoke(ContainerInterface $container,
        $requestedName, ?array $options = null)
    {
        $employeeTable = $container->get('EmployeeTable');
        $adapter = new CredentialTreatmentAdapter
            ($container->get('DbAdapter'));
        $adapter->setTableName('employees');
        $adapter->setIdentityColumn('nickname');
        $adapter->setCredentialColumn('password');

        $resourceTable = $container->get('ResourceTable');

        $encryptionMethod = array(new Employee(),
            'encrypt');

        return new IdentityManager($adapter, $resourceTable
            , $encryptionMethod);
    }
}
```

After creating several factories for models, you know that it is necessary to add an entry for this factory to the module.config.php file, in the service_manager=>factories key:

```
'IdentityManager' => Model\IdentityManagerFactory::class
```

Since `IdentityManager` has assumed responsibilities from `IndexController`, we need to refactor this controller class. Let us start with the `IndexController` constructor:

```
private ?IdentityManager $identityManager;

public function __construct(IdentityManager
    $identityManager)
{
    $this->identityManager = $identityManager;
}
```

As you can see, instead of receiving two dependencies, now `IndexController` receives only the `IdentityManager` instance.

The `IndexController->loginAction` method will be more concise with the delegation of authentication for `IdentityManager`:

```
public function loginAction()
{
    $nickname = $this->request->getPost('nickname');
    $password = $this->request->getPost('password');

    $logged = $this->identityManager->login($nickname,
        $password);

    return new JsonModel([
        'logged' => $logged
    ]);
}
```

Similarly, the `IndexController->logoutAction` method will have fewer lines of code:

```
public function logoutAction()
{
    $this->identityManager->logout();
    return $this->redirect()->toRoute('inventory');
}
```

You should think of a controller class as a manager. A manager doesn't do everything. They delegate tasks. A controller class should delegate responsibilities to model or view classes.

Of course, as the `IndexController` constructor has changed, we have to adjust the `IndexControllerFactory` class to inject an `IdentityManager` instance:

```
class IndexControllerFactory implements FactoryInterface
{
    public function __invoke(ContainerInterface $container,
        $requestedName, array $options = null)
    {
        $identityManager = $container->get
            ('IdentityManager');
        return new IndexController($identityManager);
    }
}
```

We also need to make an adjustment in the `indexAction` and `noPermissionAction` methods of the `MenuController` class. The last instruction of these methods must be as follows:

```
return new ViewModel(['user' => $auth->getIdentity()->name]);
```

This is necessary because now, the `getIdentity` method doesn't return a string value, but an object.

Finally, we need to change some lines of the `InventoryAuthorizationListener:`
`:verifyPermission` method. We have to add an `IdentityManager` instance to take care of authentication and authorization data:

```
    public static function verifyPermission
        (MvcEvent $event)
    {
        $routeName = $event->getRouteMatch()->
            getMatchedRouteName();
        if (!$routeName == 'inventory'){
            return;
        }
        $identityManager = $event->getApplication()
            ->getServiceManager()->get('IdentityManager');
        if ($identityManager->hasIdentity()){
            $params = $event->getRouteMatch()->getParams();
            $controller = $params['controller'];
            $action = $params['action'];
```

```
        if ($controller == 'menu' || $action ==
            'logout'){
            return;
        }

        $method = $event->getRequest()->getMethod();
        $permission = $controller . '.' . $action . ':'
            . $method;

        $rbac = $identityManager->getRbac();

        $role = $identityManager->getIdentity();
        if (!$rbac->isGranted($role,$permission)){
            $event->getRouteMatch()->setParam
                ('controller', 'menu');
            $event->getRouteMatch()->setParam('action',
                'no-permission');
            error_log("$role has not permission to
                $permission");
        }
    }
}
```

Now, Mr. Barnaby or other fired employees won't cause damage to the system (since a manager has changed their permissions before the dismissal).

With that, we have finally finished this chapter.

Summary

In this chapter, we have learned how to implement roles and resources registration, in a similar way to what we did in previous chapters. We talked about the concepts of roles and resources related to our project. We also learned how to associate resources with roles and how to associate roles with `Inventory` module users (employees). After presenting these registrations, we learned how to implement access control with the `laminas-permissions` component and how to verify user permissions with the aid of `laminas-eventmanager`. Finally, we saw how to decouple login and logout logic from the controller layer, creating an identity management class. We also saw how to prevent attacks by unhappy fired employees.

In the next chapter, we will learn how to implement a product basket, which will allow the `Store` module user (customer) to select products and make a purchase.

Further reading

If you want to learn more about authorization approaches, here are some good articles to read:

- ACL approach: article by Ravi S. Sandhu and Pierangela Samarati at `https://www.profsandhu.com/cs5323_s18/SS-1994.pdf`

- RBAC approach: article by Ravi S. Sandhu, Edward J. Coyne, Hal L. Feinstein, and Charles E. Youman at `https://csrc.nist.gov/CSRC/media/Projects/Role-Based-Access-Control/documents/sandhu96.pdf`

Part 3:
Review and Refactoring

In this part, you will acquire the knowledge to complete and improve some features of our sample application. You will also reflect on some of the required details in commercial web applications and implement them.

This section comprises the following chapters:

- *Chapter 11, Implementing a Product Basket*
- *Chapter 12, Reviewing and Improving Our App*
- *Chapter 13, Tips and Tricks*
- *Chapter 14, Last Considerations*

Implementing a Product Basket

This chapter covers the implementation of a product basket, an essential component of a virtual store. In *Chapter 9, Event-Driven Authentication*, we learned about the `laminas-session` component and how it can be used to manage sessions and persist data between HTTP requests. In this chapter, we will use `laminas-session` to implement a product basket.

We will first start by learning how to implement inventory management. Next, we will understand how to refactor the customer home page with view helpers. After this, we will implement a product basket. Lastly, we will learn how to make a purchase order for an authenticated customer.

At the end of this chapter, you will be able to select products from a listing page, add these products to a basket, remove products from this basket, and make a purchase order. This purchase process includes customer registration, which is very similar to the employee registration we implemented in *Chapter 9, Event-Driven Authentication*.

In this chapter, we'll be covering the following topics:

- Managing the product inventory
- Refactoring the customer home page
- Controlling the product basket
- Controlling the customers
- Decoupling the generic behavior of authentication

Technical requirements

All the code related to this chapter can be found at `https://github.com/PacktPublishing/PHP-Web-Development-with-Laminas/tree/main/chapter11/whatstore`.

Managing the product inventory

To sell products in our e-commerce system, we need to have them in our inventory. It is because of this that we need to be able to manage the inventory and the adding and removal of amounts of products.

In fact, imagine you already have implemented a RU for inventory. You may be asking, "RU? What is this?" Oh, sorry, RU stands for **Recover and Update**. "But wouldn't it be CRUD?" No, because inventory depends on products. When a product is created, we must create an inventory record for this product. When a product is removed, the related inventory record must be removed too. Thus, CRUD does not make sense when it comes to inventory, and we only need to implement recovery and updates for an inventory. Right, now that you understand it, let us explain the implementation of inventory management.

We won't repeat the details about registration that we covered in previous chapters. The implementation of inventory management is available in the source code folder of this chapter, mentioned in the *Technical requirements* section. Therefore, we will focus here only on the differences related to inventory management. The exclusive data for inventory management is based on the `inventory` table.

So, let us assume that there are the following classes in the `module/Inventory/src/Model` folder:

- `Inventory`
- `InventoryTable`
- `InventoryTableFactory`

It is important to remember that the factories of each table gateway class must be configured by the `service_manager` key in the `modules.config.php` file.

> **Important note – automated tests**
>
> You will find test classes for every controller class built in this chapter in the GitHub repository mentioned in the *Technical requirements* section. However, from now on, we won't discuss these test classes to avoid unnecessary repetitions. By now, you must be aware that tests are necessary and that it is important that you make them in your development process. From now on, we will assume that when we are creating a new controller class, the respective test class was created beforehand.

We are also assuming that there are the following classes in the `module/Inventory/src/Controller` folder:

- `InventoryAPIController`
- `InventoryAPIControllerFactory`
- `InventoryController`
- `InventoryControllerFactory`

Note that we have two pairs of classes: an API controller and its factory, and a web interface controller and its factory. It is exactly the same structure that we used for product controller layer implementation in *Chapter 7, Request Control and Data View*. These controller classes are tested by the following classes in the `module/Inventory/test/Controller` folder:

- `InventoryAPIControllerTest`
- `InventoryControllerTest`

It is important to remember that the aliases and factories of each controller class must be configured by the `controllers` key in the `modules.config.php` file.

We also have a view layer implementation for roles that is very similar to what we had made for products. Therefore, let us assume that the following view scripts are in the `module/Inventory/view/inventory/inventory` folder:

- `edit.js`
- `edit.phtml`
- `index.js`
- `index.phtml`

Here, we have two pairs of files that generate two web pages: the updating page and the listing page. The `.phtml` files have a combination of HTML and PHP code, and the `.js` files have JavaScript code exclusively.

We also have an `InventoryForm` class in the `src/Form` folder to render the inventory management form.

We need to access inventory management from the `inventory` module menu. Then, we have to add a line to the unordered list of `index.phtml`, managed by the `MenuController` class. The following line stays between the link for employee registration and the link for role registration:

```
<li><a href="<?=$this->url('inventory', [
    'controller' => 'inventory'
])?>">Inventory Management</a></li>
```

This line will be rendered by the browser as the page shown in *Figure 11.1*:

Inventory Module

Hi, MARK SPEKTOR

- Product Registration
- Discount Registration
- Employee Registration
- Inventory Management
- Role Registration
- Resource Registration
- Resources of a role
- Roles of an employee

Logout

Figure 11.1: The Inventory Module menu page with a link to inventory management

When you click on the **Inventory Management** link, you will be taken to the following page:

Whatstore   Home

Product Inventories

code	name	amount	maximum	minimum	reserved
1	ENCHANTED HAMMER	2	2	1	0
2	POWER RING	7200	7200	1	0
3	UTILITY BELT	1000	5000	1	0
4	VIBRANIUM SHIELD	1	1	1	0
5	LASSO OF TRUTH	3	3000	1	0
6	FLUX CAPACITOR	3	9999	1	0
7	TOWEL	42	1001	1	0
8	LIGHT SABER	500	10000	1	0
9	GLAIVE	1	1	1	0

Go back

© 2022 Powered by Sheat Software

Figure 11.2: A list of product inventories

The page titled **Product Inventories** shows a list of product inventories. In this case, we have nine products. We can see, for example, that there are **2** units of Enchanted Hammer. and this amount is between the maximum (**2**) and minimum (**1**). There are no reserved units. If you click on **1** in the minimum column, you will be taken to the following page:

Figure 11.3: The Inventory Management page for a product

The page titled **Inventory Management** shows a form that allows some operations on the inventory of the selected product. On this page, you can add or remove units of the product, you can set the maximum or minimum amounts, and you also can reserve an amount. The selected operation will be applied when you click on the **save** button.

However, where did these inventory records come from? There is no link to add an inventory on the **Product Inventories** page. Have you forgotten what we told you at the beginning of this section? The inventory of a product is created when the product is created. In the same way, the inventory of a product is deleted when a product is deleted. Then, these inventory records appeared when Enchanted Hammer and Power Ring were registered. For this, we had to implement the `ProductTable->save` method, overriding the inherited method from the `AbstractTable` class (from the `Generic` module). So, the `ProductTable->save` method looks like this:

```
public function save(AbstractModel $model): bool
{
    $set = $model->toArray();
    $keyName = $this->keyName;
    try {
        if (empty($model->$keyName)) {
```

```
                unset($set[$keyName]);
                $this->tableGateway->insert($set);
                $lastCode = $this->tableGateway->
                    getLastInsertValue();
                $inventory = new Inventory();
                $inventory->exchangeArray(['code_product'
                    => $lastCode]);
                $this->inventoryTable->save($inventory);
            } else {
                $this->tableGateway->update($set, [
                    $keyName => $set[$keyName]
                ]);
            }
        } catch (\Exception $e) {
            error_log($e->getMessage());
            return false;
        }
        return true;
    }
```

Note that after having included a product in the `products` table, we use the `getLastInsertValue` method to recover the generated product code. Then, we insert a record into the `inventory` table using this product code. So, for each inserted product, one inventory is created.

As you can see, we use an `InventoryTable` instance to persist data into the `inventory` table. For this, we had to add an attribute for the inventory table gateway and change the `ProductTable` constructor method to receive an `InventoryTable` instance:

```
private ?InventoryTable $inventoryTable;
public function __construct(TableGatewayInterface
    $tableGateway, InventoryTable $inventoryTable)
{
    $this->tableGateway = $tableGateway;
    $this->inventoryTable = $inventoryTable;
}
```

In sequence, we show the __invoke method of `ProductTableFactory`, which also needs to be modified to make the injection of an `InventoryTable` instance:

```
public function __invoke(ContainerInterface $container,
    $requestedName, ?array $options = null)
{
    $adapter = $container->get('DbAdapter');
    $resultSetPrototype = new ResultSet();
    $resultSetPrototype->setArrayObjectPrototype(new
        Product());
    $tableGateway = new TableGateway('products',
        $adapter, null, $resultSetPrototype);
    $inventoryTable = $container->
        get('InventoryTable');
    $productTable = new ProductTable($tableGateway,
        $inventoryTable);
    return $productTable;
}
```

As previously mentioned, when we remove a product from registration, we need to delete the related inventory record. So, the `ProductTable->delete` method must look like this:

```
public function delete($value): bool
{
    try {
        $this->inventoryTable->delete([
            'code_product' => $value
        ]);
        $this->tableGateway->delete([
            $this->keyName => $value
        ]);
    } catch (\Exception $e) {
        error_log($e->getMessage());
        return false;
    }
    return true;
}
```

As you can see, we deleted the dependent record before because there is no inventory without the product.

Let us come back to the page titled **Inventory Management** (*Figure 11.3*). When the user clicks on the **save** button on the form, a request is processed by the update method of the InventoryAPIController class. This method calls a specific method of the InventoryTable class, according to the selected operation. The source code of InventoryAPIController->update looks like this:

```php
public function update($id, $data)
{
    $inventory = $this->inventoryTable->
        getByField('code', $id);
    $operation = $data['operation'];
    $methods = [
        'add' => 'addItems',
        'remove' => 'subtractItems',
        'reserve' => 'reserveItems',
        'maximum' => 'setMaximum',
        'minimum' => 'setMinimum'
    ];
    $method = ($methods[$operation] ?? 'invalid');
    if ($method == 'invalid'){
        $updated = false;
    } else {
        $updated = $this->inventoryTable->$method
            ($id,$data['amount']);
    }
    $updated = (bool) $updated;
    return new JsonModel(['updated' => $updated]);
}
```

> **Variable functions**
>
> Instead of using a switch structure, the InventoryAPIController->update method uses the **variable function** feature of PHP. A variable function is a function (or method) defined by a variable. In the update method, we use the $method variable to call a method from the InventoryTable instance. The method to be called will be defined at runtime. Variable functions are very useful to write cleaner code.

These are the specific implementations of inventory management. Then, the inventory management is ready to use. In the next section, we will refactor the customer home page and change the default routing of our application.

Refactoring the customer home page

In this section, we will create the home page for the `Store` module, which is the module for customers of whatstore. For this, we will modify the route configuration, and the `IndexController` class of the `Store` module, as well as view script files.

The general home page of whatstore application will be the `Store` module home page. Thus, we will shift the home route from the `Application` module to the `Store` module. For this, we must copy the following part of the `module.config.php` file of the `Application` module:

```php
'home' => [
    'type'    => Literal::class,
    'options' => [
        'route'    => '/',
        'defaults' => [
            'controller' => Controller\
                IndexController::class,
            'action'     => 'index',
        ],
    ],
],
```

This part must be pasted into the `module.config.php` file of the `Store` module. So, the index page of the `Store` module becomes the customer home page. This new home page will present the store products to the customer. We can have only one route name for an application, and then you need to delete the home route block code of the `module.config.php` file in the `Application` module.

To present the products, the `IndexController` class must have a table gateway for products. This table gateway is ready; it is the `ProductTable` class. We only need to inject an instance of `ProductTable` into the `IndexController` instance. You should recall how to do this because we did something similar in *Chapter 7, Request Control and Data View*. Let us focus on the `IndexController` implementation. In sequence, we can see the content of this class (except for the namespace declarations):

```php
class IndexController extends AbstractActionController
{
    private ProductTable $productTable;
```

```
public function __construct(ProductTable $productTable)
{
    $this->productTable = $productTable;
}
public function indexAction()
{
    $products = $this->productTable->getAll();
    $form = new ProductForm();
    return new ViewModel([
        'form' => $form,
        'products' => $products
    ]);
}
}
```

The IndexController constructor method receives an instance of the ProductTable class. The ProductTable instance is injected by the IndexControllerFactory class. This factoring, as we know after creating several controllers, is defined in the module.config.php file.

The IndexControlller->indexAction method recovers a product set from the products table and creates an object from the ProductForm class. The product set and the form object are delivered to the layer view through the ViewModel class. As you now know, this means that these two objects ($form and $products) can be handled by an index.phtml file. Let us see and understand the content of this view file step by step:

1. First, we declare the namespace for AuthenticationService:

    ```
    <?php
    use Laminas\Authentication\AuthenticationService;
    ```

    This is necessary because we will control the customer authentication, enabling the appropriate links according to the customer state.

2. We add a documentation block, declaring a type for the $this keyword:

    ```
    /**
     * @var Laminas\View\Renderer\PhpRenderer $this
     */
    ```

    This is beneficial for the IDEs because there is no possibility for an IDE to infer what $this is out of a class. At runtime, the index.phtml file will be processed inside a class, and $this will make sense to the PHP interpreter.

3. We import a JavaScript file to handle the client-side events for the HTML page produced by `index.phtml` and `layout.phtml`:

```
$this->headScript()->appendScript
    (file_get_contents(__DIR__ . '/index.js'));
```

4. In this view file, we use a helper class named `PartialLoop`. This class replaces the `foreach` structure and allows us to use a template file for repeated partial views. The `PartialLoop` instance is recovered through the `$this->partialLoop` method. `PartialLoop` directly receives a collection of objects, but sometimes, we need to use additional variables. One way to share variables between controller view scripts and helper view scripts is to store them in a `Placeholder` helper. In this case, we need to store a form for each product and an index to control the number of exhibited products by line:

```
$this->placeholder('products')->form = $this->form;
$this->placeholder('products')->index = 1;
```

5. When we use the `foreach` structure, we define a variable to receive one element of a collection for each iteration. `PartialLoop` allows us to define this element variable through the `setObjectKey` method:

```
$this->partialLoop()->setObjectKey('product');
```

6. We use an `AuthenticationService` instance to check whether the current customer is authenticated. If the customer is authenticated, a **Logout** link will be shown; otherwise, a **Login** link will be shown:

```
$authenticationService = new AuthenticationService();
$link = ['action' => 'index', 'label' => 'Login'];
if ($authenticationService->hasIdentity()){
    $link = ['action' => 'logout', 'label' =>
        'Logout'];
}
?>
<h1>Our products</h1>
<table style="width: 100%">
<tbody>
<tr>
<td><a href="<?=$this->url('store',['controller' =>
    'productbasket'])?>">Product Basket</a></td>
<td><a href="<?=$this->url('store',['controller' =>
```

```
            'customer','action' => $link['action']])?>">
                <?=$link['label']?></a></td>
</tr>
</tbody>
</table>
```

7. Finally, we use PartialLoop to generate the view for the product set and define hidden tags to store URLs to feed the JavaScript functions:

```
<?=$this->partialLoop('products',$this->products)?>
<span id="api" url="<?=$this->url('storeapi',
    ['controller' => 'productbasketapi'])?>"></span>
<span id="target" url="<?=$this->url('store',
    ['controller' => 'productbasket'])?>"></span>
```

Note that the PartialLoop helper receives two arguments. The second argument is the product set, recovered by ProductTable and sent by IndexController. But what is the first argument? The first argument is the name of a view template file. The templates related to PartialLoop must stay in the root of the view folder. So, we need to create the products.phtml file in the module/ Store/view folder. Let us look at the source code of products.phtml part by part:

1. This file starts with a decision structure that controls the opening of div with a class attribute equal to "row". This ensures that each row has three products. We use alternative syntax for the if structure because this syntax is clearer when we have HTML code interleaved with PHP code:

```
<?php
$index = $this->placeholder('products')->index;
if ((($index - 1) % 3) == 0):
?>
<div class="row">
<?php
endif;
?>
```

2. Each product is presented in an HTML format, within a div tag that is configured as a column for a row with three columns. The amount to be requested for each product is set to 1 by default:

```
<div class="col-md-3">
    <p>
<?php
```

```php
        $form = $this->placeholder('products')->form;
        $form->setAttribute('id', 'product' . $this-
            >product->code);
        $form->bind($this->product);
        $form->get('amount')->setValue(1);
        echo $this->form()->openTag($form);
        echo $this->formText($form->get('name')) .
            '<br/>';
        echo $this->formLabel($form->get('price'));
        echo $this->formText($form->get('price')) .
            '<br/>';
        echo $this->formLabel($form->get('amount'));
        echo $this->formNumber($form->get('amount')) .
            '<br/>';
        echo $this->formHidden($form->get('code'));
        echo $this->form()->closeTag();
?>
    </p>
    <button class="buy" id="<?=$this->product->code?>">
        Buy</button>
</div>
```

3. Finally, we check whether there are three products inside the div tag to close it. For each product, we increment the $index variable:

```php
<?php
        $index = $this->placeholder('products')->index;
        if ($index % 3 == 0):
?>
</div>
<?php
        endif;
        $this->placeholder('products')->index = ++$index;
?>
```

Now, when you make a request to `localhost:8000` in your browser, the resulting page will look like *Figure 11.4*:

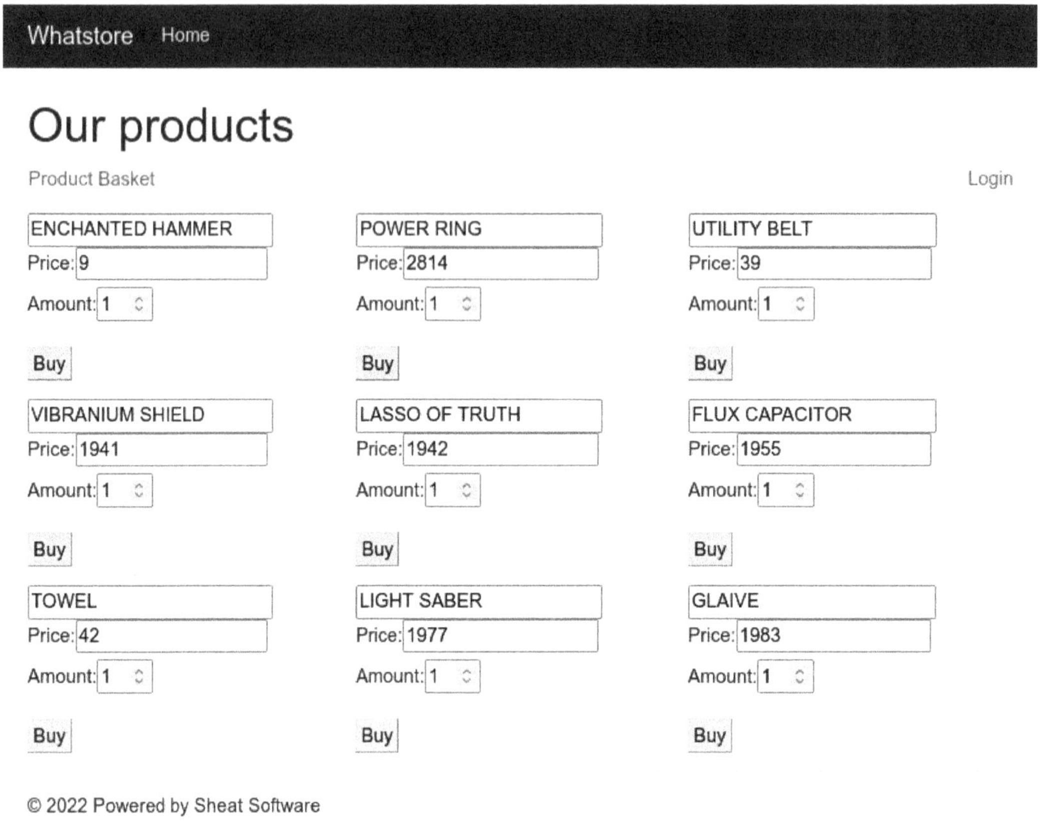

Figure 11.4: The customer home page with available products

As you can see, every product on the customer home page has a **Buy** button. When a customer clicks on this button, the product must be added to the product basket. This action is controlled by the `index.js` file, located in the `view/store/index` folder:

index.js

```
var addToBasket = function(){
    var api = $( "#api" ).attr("url") + '/' + $(this).
        attr('id');
    var formData = $( "#product" + $(this).attr('id')
        ).serialize();
```

```
$.post(api, formData).done(function(data){
    var target = $( "#target" ).attr("url");
    document.location = target;
}).fail(function(){
    alert('error to insert product into basket');
});
};


$(document).ready(function(){
    $("button[class='buy']").click(addToBasket);
});
```

The click event of the **Buy** button triggers the addToBasket function, which sends an HTTP POST request to the ProductBasketAPIController class.

In the next section, we will implement this class and others related to the product basket.

Controlling the product basket

In this section, we will implement code that provides control of the product basket for the customer.

When a customer clicks on the **Buy** button of a product on the customer home page, the request is handled by the ProductBasketAPIController->create method:

```
public function create($data)
{
    $inserted = $this->productBasket->create($data);
    return new JsonModel(['inserted' => $inserted]);
}
```

As you can see, the ProductBasketAPIController->create method calls the ProductBasket->create method.

The ProductBasket->create method, the source code of which we can see in sequence, adds the product data to a container object if it's doing it for the first time and increments the requested amount of this product each time thereafter:

```
public function create(array $data): bool
{
    $inserted = false;
```

```
    if (isset($this->container->productBasket[$data
        ['code']])){
        $this->container->productBasket
        [$data['code']]['amount'] += $data['amount'];
        $inserted = true;
    } else {
        $this->container->productBasket[$data
            ['code']] = $data;
        $inserted = true;
    }
    return $inserted;
}
```

The container object stores an array. This array is initialized in the ProductBasket class constructor:

```
public function __construct(Container $container)
{
    $this->container = $container;
    if (!isset($this->container->productBasket)){
        $this->container->productBasket = [];
    }
}
```

The container object is an instance of Laminas\Session\Container. We already used this class in *Chapter 9, Event-Driven Authentication*, to store an employee identity in session. More precisely, the identity was stored in a Laminas\Authentication\Storage\Session object that uses a Laminas\Session\Container object internally for storage and persistence when the session manager writes the closed session. The developer does nothing to store this data other than calling the authenticate method.

After adding a product to the product basket, index.js redirects the customer to the localhost:8000/store/productbasket endpoint. This address will render a page that looks like this:

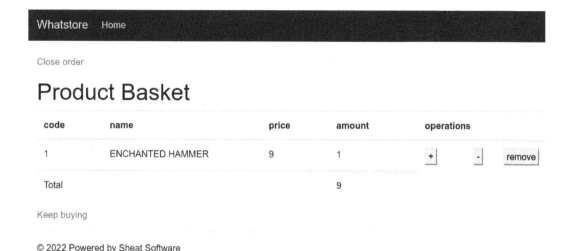

Figure 11.5: The content of the product basket

The page of the product basket is produced by the `view/store/product-basket/index.phtml` file. This file receives the output of the `ProductBasket->getProducts` method, which is invoked by the `ProductBasketController->indexAction` method. You can increment or decrement the amount of a product by clicking on the + or – buttons in the **operations** column. To remove a product from the basket, you click on the **remove** button.

When you click on the + or – button, the `ProductBasketAPIController->update` method calls the `ProductBasket->update` method, and the following is its source code:

```
public function update(int $id, array $data): bool
{
    $updated = false;
    if (!isset($this->container->productBasket[$id]))
    {
        return $updated;
    }
    $operation = ($data['operation'] ?? 'invalid');
    if ($operation == 'add'){
        $this->container->productBasket[$id]
            ['amount']++;
        $updated = true;
    }
    if ($operation == 'sub' && ($this->container->
```

```
            productBasket[$id]['amount'] > 1)){
        $this->container->productBasket[$id]
            ['amount']--;
        $updated = true;
    }
    return $updated;
}
```

The `ProductBasket->update` method increments or decrements, depending on the selected operation. This method ensures that the minimum amount is **1**.

When you click on the **remove** button, the `ProductBasketAPIController->delete` method calls the `ProductBasket->delete` method, and the following is its source code:

```
public function delete(int $id): bool
{
    if (!isset($this->container->productBasket[$id])){
        return false;
    }
    unset($this->container->productBasket[$id]);
    return true;
}
```

After adding all the desired products to the basket, you can make a purchase by clicking on the **Close order** link. After clicking on this link, you should see a page like this:

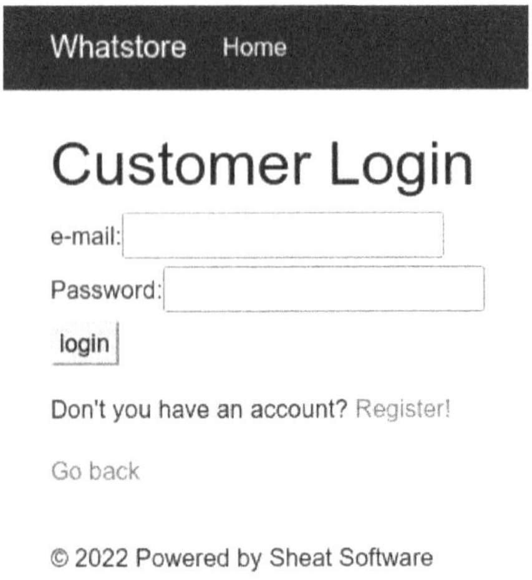

Figure 11.6: The login page for customers

As you can see, *Figure 11.6* shows a login form for the customer. This happens because the customer needs to authenticate to finish the purchase. In the next section, we will talk about the registering and authentication of customers.

Controlling the customers

As discussed in the previous section, a customer needs to be authenticated to finish a purchase. Of course, at first, the customer will not have an account. It is because of this that there is a **Register** link on the login page. If you click on this link, you will see a page looking like this:

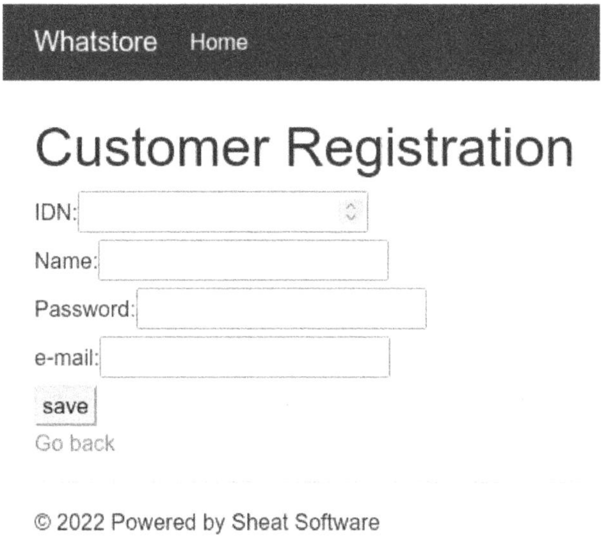

Figure 11.7: The registration page for customers

We won't repeat the details about previous registrations. We will assume that there are the following classes in the `module/Store/src/Model` folder:

- `Customer`
- `CustomerTable`
- `CustomerTableFactory`

Never forget that the factories of each table gateway class must be configured by the `service_manager` key in the `modules.config.php` file.

We also assume that the following classes are in the `module/Store/src/Controller` folder:

- `CustomerAPIController`
- `CustomerAPIControllerFactory`
- `CustomerController`
- `CustomerControllerFactory`

There is a difference in customer registration compared to previous registrations. The API only allows customer creation. It is not possible to update or delete a customer. Of course, this is to avoid a customer removing another customer or changing the personal data of another customer. The customer controller classes are tested by the following classes in the `module/Store/test/Controller` folder:

- `CustomerAPIControllerTest`
- `CustomerControllerTest`

Never forget that aliases and factories of each controller class must be configured by the `controllers` key in the `modules.config.php` file.

> **Do controllers always require aliases?**
>
> No, controllers aliases are not required if controllers are not part of route paths. However, as we receive controller aliases as parameters in our sample application, we need to define these aliases.

To control customer authentication, we use the `StoreAuthenticationListener` class, which is very similar to the `InventoryAuthenticationListener` class. Let us see the `StoreAuthenticationListener::verifyIdentity` method:

```php
public static function verifyIdentity(MvcEvent $event)
{
    $routeName = $event->getRouteMatch()->
        getMatchedRouteName();
    if ($routeName !== 'store'){
        return;
    }
    $params = $event->getRouteMatch()->getParams();
    $controller = $params['controller'];
    if ($controller !== 'order'){
        return;
    }
    $authenticationService = new AuthenticationS
```

```
        ervice();
    if (!$authenticationService->hasIdentity()){
        $action = $params['action'];
        if ($controller == 'index' && ($action ==
            'index' || $action == 'login')){
            return;
        }
        $event->getRouteMatch()->setParam
            ('controller', 'customer');
        $event->getRouteMatch()->setParam('action',
            'index');
    }
}
```

Note that this StoreAuthenticationListener class only controls the requests for the Store module, in the same way that the InventoryAuthenticationListener class only controls the requests for the Inventory module. Of course, for the StoreAuthenticationListener class to listen to the route event, we need to attach it to ServiceManager in the Module.php file of the Store module, as we did in the Inventory module. We need to add the onBootstrap method to Module class, as you can see in the next snippet:

```
public function onBootstrap($e)
{
    if (defined('PHPUNIT_COMPOSER_INSTALL')) return;
    $application = $e->getApplication();
    $application->getEventManager()>attach
    (MvcEvent::EVENT_ROUTE, [StoreAuthenticationListener
        ::class,'verifyIdentity']);
}
```

We also have a view layer implementation for customers that is very similar to what we had made for products and other models. So, we are assuming that the following view scripts are in the module/Store/view/store/customer folder:

- index.js
- index.phtml
- register.js
- register.phtml

Here, we have two pairs of files that generate two web pages: the login page (`index.phtml`) and the registration page (`register.phtml`).

After registering, the customer can log in. The authentication for the customer is very similar to what we did for the employee in the `Inventory` module. One significant difference is that we store a `Customer` object in a session. In the `CustomerController->loginAction` method, we recover the `Customer` object using the authentication adapter:

```
Customer::setCustomer($this->adapter->getResultRowObject(null, '
password'));
```

The `AbstractAdapter->getResultRowObject` method has two mutually exclusive arguments. The first argument defines which fields from the table you want to recover. The second argument defines which fields from the table you want to omit from recovery.

The `Customer::setCustomer` static method stores the `Customer` instance in the session:

```
public static function setCustomer(object $customer): void
{
    $container = new Container();
    $container->customer = $customer;
}
```

Why do we need to store the `Customer` object in the session? Don't we get the identity data with `AuthenticationService`? Yes, `AuthenticationService` stores the identity used for authentication, which is, for the customer, the email address. But `AuthenticationService` does not automatically store other data. If we want to show the customer's name on any page, we have to store it explicitly.

After login, the customer is redirected to the page to close the purchase order, as we can see in *Figure 11.8*:

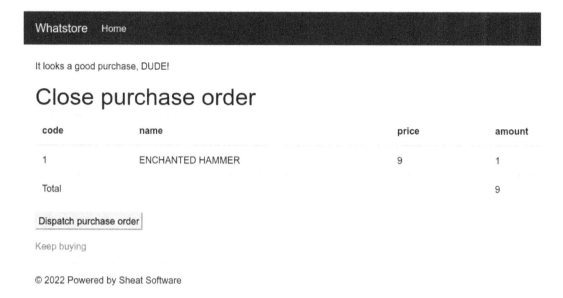

Figure 11.8: The page to close the purchase order

You can see that the message at the top has the customer's name. This value is recovered from the session with Customer::getCustomer. In the next section, we will understand the code behind the page for closing a purchase order.

Order closing

In the previous section, we saw the customer redirected to the localhost:8000/store/order endpoint. To conclude the purchase, the customer must click on the **Dispatch purchase order** button. This action triggers the closeOrder function, contained in the index.js file:

index.js

```
var closeOrder = function(){
  var api = $( "#api" ).attr("url") + '/0';
  var target = $( "#target" ).attr("url");
  $.post(api).done(function(data){
    if (!data.inserted)
    {
      alert('purchase order has been not inserted!');
      return;
    }
```

```
      alert('Thank you for buying in Whatstore!
            Come back always!')
      document.location = target;
    }).fail(function(){
      alert('error to close the purchase order');
    });
  };


  $(document).ready(function(){
    $("#close").click(closeOrder);
  });
```

The `closeOrder` function sends an HTTP POST request to the `localhost:8000/store/orderapi` endpoint. This means the execution of the `OrderAPIController->create` method. Let us see the source code of this method, part by part:

1. First, we recover the **identity document number (IDN)** of the customer and define the initial state for the purchase order:

    ```
    public function create($data)
    {
        $idn = Customer::getCustomer()->IDN;

        $set = [
            'status' => PurchaseOrder::CREATED,
            'IDN' => $idn
        ];
    ```

2. Then, we insert a record into the `purchase_orders` table. After inserting, we recover the generated value for the code field:

    ```
    $inserted = false;
    try {
        $purchaseOrder = new PurchaseOrder();
        $purchaseOrder->exchangeArray($set);
        $this->purchaseOrderTable->save
            ($purchaseOrder);
        $code = $this->purchaseOrderTable->
            getLastCreatedCode();
    ```

3. In sequence, we iterate over the product basket and insert one record into the order_items table for each product selected by the customer:

```php
$products = $this->productBasket->
    getProducts();

foreach($products as $product){
    $set = [
        'code_order' => $code,
        'code_product' => $product
            ['code'],
        'price' => $product['price'],
        'amount' => $product['amount']
    ];
    $orderItem = new OrderItem();
    $orderItem->exchangeArray($set);
    $this->orderItemTable->save
        ($orderItem);
}
$this->productBasket->clear();
$inserted = true;
} catch (\Exception $e) {
    error_log($e->getMessage());
}

return new JsonModel(['inserted' => $inserted]);
}
```

As you can see, after inserting the purchase order and its items, the OrderAPIController->create method cleans the product basket. At the end of this method, the response is sent to the closeOrder function of the index.js file. If everything works, the customer will see the following message:

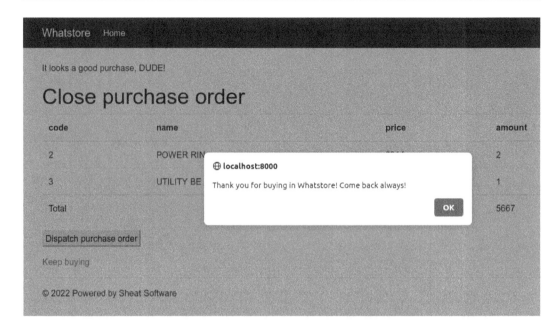

Figure 11.9: The message for a successful purchase

After clicking on the **OK** button, the customer will be redirected to the product basket page, which is void now because the product basket was cleared.

But, wait! What about the payment? And the monitoring of product delivery? Yes, yes, these are real and necessary features for an e-commerce system, but we won't implement these features in this book. Remember that our focus is on the Laminas components' features. The whatstore project is our tool to learn Laminas. Currently, we have enough implementations to understand the components used so far.

> **Payment components for PHP applications**
>
> You can find PHP payment components in repositories, such as Packagist (`https://packagist.org`), and in code repositories of payment integrators, such as PayPal (`https://github.com/paypal`).

As proposed at the beginning of this chapter, we have implemented a product basket, including updating and deleting products from it and reading it to create a purchase order. Finally, we will show you how to improve identity management for the `Inventory` and `Store` modules by decoupling the generic behavior between them.

Decoupling the generic behavior of authentication

We know that the `Inventory` and `Store` modules have authentication, but they authenticate against different sources. However, it is possible to isolate what is common for both modules into a generic management identity class.

We will extract the generic code from `Inventory\Model\IdentityManager` to `Generic\Model\IdentityManager`, in our reuse module, `Generic`. Let us talk only the things that will be different in new classes.

First, the generic `IdentityManager` class will have three protected attributes. This class doesn't have an attribute for the resource table gateway because it is a particular case for the `Inventory` module:

```
protected ?AdapterInterface $adapter;
protected ?array $encryptionMethod;
protected ?AuthenticationService $auth;

public function __construct(AdapterInterface $adapter,
    array $encryptionMethod)
{

    $this->adapter = $adapter;
    $this->encryptionMethod = $encryptionMethod;
    $this->auth = new AuthenticationService();

}
```

Second, the `IdentityManager->login` method calls a `doCustomTasks` method if the authentication result is valid. The `doCustomTasks` method is declared (it is not abstract), but it is void. `Generic\Model\IdentityManager` is not an abstract class, so it can be instantiated directly if it is implemented is enough. This feature aims to avoid unnecessary inheritance.

After creating `Generic\Model\IdentityManager`, we modify `Inventory\Model\IdentityManager` to inherit from the first class. So, the new `Inventory\Model\IdentityManager` is reduced to four methods. The first method is the setter for the resource table gateway:

```
private ?ResourceTable $resourceTable;

public function setResourceTable(ResourceTable
    $resourceTable)
{

    $this->resourceTable = $resourceTable;

}
```

The second method is the overriding of the `doCustomTasks` method, to write the `Employee` instance into the session without a password and loading the access control data:

```
protected function doCustomTasks(): void
{
    $this->auth->getStorage()->write($this->adapter->
        getResultRowObject(null,'password'));
    $this->loadAuthorizationData();
}
```

The other two methods, `loadAuthorizationData` and `getRbac`, keep the same implementation, so there is no need to show them. However, the `__invoke` method of `Inventory\Model\IdentityManagerFactory` needs to be modified because the dependency injection is made now in two stages, first with the `IdentityManager` constructor and then with the `setResourceTable` method:

```
public function __invoke(ContainerInterface $container,
$requestedName, ?array $options = null)
{
    $adapter = new CredentialTreatmentAdapter
        ($container->get('DbAdapter'));
    $adapter->setTableName('employees');
    $adapter->setIdentityColumn('nickname');
    $adapter->setCredentialColumn('password');

    $resourceTable = $container->get('ResourceTable');

    $encryptionMethod = array(new Employee(),
        'encrypt');

    $identityManager = new IdentityManager($adapter,
        $encryptionMethod);
    $identityManager->setResourceTable($resourceTable);
    return $identityManager;
}
```

After extending `Generic\Model\IdentityManager` for the `Inventory` module, we do the same for the `Store` module. `Store\Model\IdentityManager` is simpler than its counterpart in the `Inventory` module. There is only one method, `doCustomTasks`:

```php
<?php
namespace Store\Model;

use Generic\Model\IdentityManager as
    GenericIdentityManager;

class IdentityManager extends GenericIdentityManager
{
    protected function doCustomTasks(): void
    {
        Customer::setCustomer($this->adapter->
            getResultRowObject(null, 'password'));
    }
}
```

Of course, as this class has dependencies, it needs a factory class. The `IdentityManagerFactory->__invoke` method should have this content:

```php
    public function __invoke(ContainerInterface $container,
        $requestedName, ?array $options = null)
    {
        $adapter = new CredentialTreatmentAdapter
            ($container->get('DbAdapter'));
        $adapter->setTableName('customers');
        $adapter->setIdentityColumn('email');
        $adapter->setCredentialColumn('password');

        $encryptionMethod = array(new Customer(),
            'encrypt');

        return new IdentityManager($adapter,
            $encryptionMethod);
    }
```

As we know, a factory class needs to be defined in the `module.config.php` file of the module. Then, we add the following item to the `service_manager=>factories` key:

```
IdentityManager::class => Model\IdentityManagerFactory::class
```

You may ask, "Why don't you just use `IdentityManager` as the key?" The answer is that there is already an `IdentityManager` key for the `Inventory` module. So, here we have found a problem with using factory keys based on short aliases: name collision. Until now, we didn't have this problem, but in case it arises, it is a good practice to use **Full Qualified Class Names (FQCN)** as keys to avoid name collision. In PHP, an FQCN is provided by the `class` static property of the class.

Now, we can really end this chapter.

Summary

In this chapter, we learned how to implement inventory management, using the acquired knowledge from the previous chapter about registration creation. In sequence, we learned how to refactor the customer home page with `Placeholder` and `PartialLoop` view helpers. We changed the home page route of the whatstore application for the `Store` module because the default user must be the customer.

Then, we learned how to implement a product basket, using `Laminas\Session\Container` to store the products in the session. Finally, we learned how to make a purchase order for an authenticated customer.

Finally, we saw how to decouple the generic behavior of authentication for each module so that it implements only specific behavior.

In the next chapter, we will review implemented features, and we will refactor code when it is appropriate. We will also create a product search box, generate API documentation, and encapsulate identity management.

12

Reviewing and Improving Our App

This chapter is dedicated to reviewing and improving our whatstore application. In the last chapter, we delivered the requirements of the first version of the application. However, no software version lasts forever. New software is always under construction. Software that is not constantly changing is probably dead software.

As we believe that concious about software maintenance is so important, we will start this chapter by talking about fundamental agile development principles. An agile approach recognizes that software requirements can change during the execution of a project.

As software developers, we need to ensure that any changes to our code can be made as easily as possible. For this, our code needs to be as simple as possible. Because of this, in this chapter, we will talk about the concept of clean code and show how to eliminate duplicated code, which can make our code very complicated.

In this chapter, we will also make a little change in the whatstore application, as an example of adding desirable features that were discovered only at the time of delivery of a version. It is not a problem because we can always release a new version. The question is how fast we can do it. It is because of this we are using a framework and several tools and techniques.

Software can be very powerful, but if nobody knows how to use it, it is useless, or at best, it is software made only for its creators. Software needs documentation, and a part of this is API documentation. In this chapter, we will learn how to generate API documentation using OpenAPI Specification.

Finally, we will show that software needs continuous deep analysis because although it may be working, we can create unexpected behaviors for it if we don't undertake exhaustive tests. For every software, always expect the unexpected. In this chapter, we will show an example of unexpected behavior in the identity management of our application, something that is not perceivable when you only implement ordinary use cases.

In short, in this chapter, we'll be covering the following topics:

- Learning agile development principles
- Eliminating duplicated code
- Removing unnecessary configurations
- Creating a product search box
- Generating API documentation
- Isolating identity management for each module

Technical requirements

All the code related to this chapter can be found at `https://github.com/PacktPublishing/PHP-Web-Development-with-Laminas/tree/main/chapter12/whatstore`.

Principles of agile development

This chapter consists of an exercise for an agile development process with **test-driven development** (**TDD**). One of the principles of the **Agile Manifesto** says that we should pay continuous attention to technical excellence and practice good design because this enhances agility. You can read more about the Agile Manifesto at `https://agilemanifesto.org`.

The third task of the TDD cycle is refactoring. In the book *Test Driven Development by Example* by Kent Beck, his definition of refactoring is to *"eliminate all of the duplication created in merely getting the test to work."* Thus, refactoring seems only to be a simple cleaning task, but Martin Fowler warns that refactoring is fundamental to ensure code is easy to maintain. He says:

> *A poorly designed system is hard to change – because it is difficult to figure out what to change and how these changes will interact with the existing code to get the behavior I want.*

Martin Fowler, Refactoring: Improving the Design of Existing Code

Another principle of the Agile Manifesto says that *working software is the primary measure of progress*. Essentially, this is what your customer wants – something that works. However, there is another manifesto too, the **Manifesto for Software Craftsmanship**, which says that we should value *not only working software but also well-crafted software*. You can read more about the Manifesto for Software Craftsmanship at `https://manifesto.softwarecraftsmanship.org`.

From *Chapter 6, Models and Object-Relational Mappers with Behavior-Driven Development*, to *Chapter 11, Implementing a Product Basket*, we implemented the use cases described in *Chapter 5, Creating the Virtual Store Project*. We now have a functional web application, using several Laminas components. In this chapter, we will review and improve the implementation of our e-commerce application to generate cleaner code that is easier to maintain.

Remember this: you can only improve something that already exists. Code that works can be improved. Code that works also usually needs to be changed to allow new features or fix bugs. After making the required changes, code usually needs to be reviewed to avoid maintenance troubles in the future. Software improvement is an endless process. While it is useful, it is a product in a continuous improvement process.

Eliminating duplicated code

A computer program is a set of instructions. The wonderful thing about computers is that we don't need to repeat an instruction for them if this instruction is already stored. We only need to say where the instruction is, like pointing somebody who is reading a handbook to a specific page. In fact, if you give the handbook to somebody, they don't need to ask you what is written in it anymore.

When we create software, we create several handbooks. These handbooks can be procedures, functions, classes, or traits. All of these code structures are reusable units of code. We have several options to avoid duplicated lines of code, by centralizing them in one of these structures.

Duplicated lines of code can generate more work because you will have to modify more than one line, possibly in more than one file, to implement a new feature or fix a bug. For this reason, we have to avoid duplicated lines of code.

There is a useful tool to detect duplicated lines in PHP programs – **PHP Copy/Paste Detector (PHPCPD)**. You can download the latest version of PHPCPD from `https://phar.phpunit.de/phpcpd.phar`. The downloaded file is `phpcpd.phar`. Rename this file `phpcpd` and permit it to execute. Move `phpcpd` to a directory of the system path (which is visible everywhere).

The `phpcpd` program compares PHP code from a folder and shows the pairs of duplicated code blocks. Let us run `phpcpd`. However, before we do this, we need to delete the `vendor` subdirectory inside the `whatstore` directory. Why? It is because we have to analyze only our code and not the dependencies code.

So, let's run the following command:

```
$ phpcpd whatstore/
```

The output will look like this:

```
phpcpd 6.0.3 by Sebastian Bergmann.


Found 1 clones with 21 duplicated lines in 2 files:

  - /opt/lampp/htdocs/whatstore/module/Inventory/src/Model/
Discount.php:31-52 (21 lines)
```

```
/opt/lampp/htdocs/whatstore/module/Inventory/src/Model/
Product.php:33-54


0.33% duplicated lines out of 6422 total lines of code.
Average size of duplication is 21 lines, largest clone has 21
of lines


Time: 00:00.021, Memory: 4.00 MB
```

As you can see, there are a few duplicated PHP lines in our whatstore project. Since you have detected the files with duplicated lines, you can use an IDE such as Eclipse to check whether it is really necessary to refactor the source code. You can compare the two files in Eclipse by selecting both of them and right-clicking on one of them. Choose the **Compare With | Each Other** option, and Eclipse will open a window looking like this:

Figure 12.1: File comparison in Eclipse

According to the phpcpd output, we have compared the Discount.php and Product.php files. As we can see, there are some duplicated code blocks in the getInputFilter method. These code blocks are related to the addition of input fields and their filters and validators to the InputFilter instance.

This is the time to refactor these code blocks and turn them into clean code. Robert C. Martin, in his book *Clean Code*, says that **clean code** is readable code. You may ask, how do I know whether code is readable? Well, Scott Ambler, in his book *The Object Primer*, recommends that you follow the **30-second rule**: if someone else does not understand your code in less than 30 seconds, then you need to rewrite it. I have an additional suggestion: try to write code blocks that fit in the window of your text editor. If you need to scroll to see the full code of a function or method, this subroutine is a strong candidate to be shortened.

> **Source Code, Grass, and Rabbits**
>
> There is something common among source code, grass, and rabbits: they can multiply very fast. But it is easier to lose control over source code than over grass or rabbits.

Let us take an example of duplicated code reduction using the output of `phpcpd`. If you observe the implementation of the `getInputFilter` method in *Figure 12.1*, there are repeated steps for each input: input creation, filter addition, and validator additions. We can create an extension of the `Laminas\InputFilter\InputFilter` class with a method that encapsulates these steps. So, let us create the `InputFilter` class in the `Generic` module, within the `Generic\Model` namespace like so:

InputFilter.php

```php
<?php
declare(strict_types=1);


namespace Generic\Model;


use Laminas\InputFilter\InputFilter as LaminasInputFilter;
use Laminas\InputFilter\Input;
use Laminas\Filter\FilterChain;
use Laminas\Validator\ValidatorChain;


class InputFilter extends LaminasInputFilter
{
    public function addInput(string $name, array $filters = [],
array $validators = []): self
    {
        $input = new Input($name);
        $filterChain = new FilterChain();
        foreach ($filters as $filter) {
            $filterChain->attach($filter);
        }
        $input->setFilterChain($filterChain);
```

```
        $validatorChain = new ValidatorChain();
        foreach($validators as $validator){
            $validatorChain->attach($validator);
        }
        $input->setValidatorChain($validatorChain);
        $this->add($input);
        return $this;
    }
}
```

The `addInput` method receives the input name, a filter set, and a validator set as arguments. The `addInput` method returns a reference for the `InputFilter` instance. So, we can call more than one method from the `InputFilter` instance in one only instruction – for example, to add two fields in one only instruction:

```
$inputFilter = new InputFilter();
$inputFilter->addInput("code")->addInput("name");
```

You don't need to repeat the variable name of the `InputFilter` instance to call another method, or to call the same method several times. It is an advantage to return `$this` for methods that wouldn't return anything.

Next, we will replace `Laminas\InputFilter\InputFilter` with `Generic\Model\InputFilter` in the `Discount` and `Product` classes. This replacement will reduce the `getInputFilter` implementations in some lines. For example, the `getInputFilter` method of the `Discount` class will look like this:

```
public function getInputFilter(): InputFilter
{
    $inputFilter = new InputFilter();

    $inputFilter->addInput('code', [new ToInt()])
    ->addInput('name',
        [
            new StringToUpper(),
            new Alnum(true)
        ],
        [new StringLength(['min' => 3])]
    )
```

```
        ->addInput('operator', [
            new AllowList(['list' => ['-','*','/']])
            ]
        )
        ->addInput('factor', [new ToFloat()]);

        return $inputFilter;
    }
```

Observe that we call the addInput method several times, but we need to use the $inputFilter variable only once. It is because the addInput method returns the reference to the InputFilter instance.

Well, we have reduced the Discount->getInputFilter method by several lines. In the same way, we can reduce the Product->getInputFilter method by several lines. After refactoring the source code, we need to run tests to check whether everything keeps working. Before running the tests, we need to recreate the vendor directory by running the composer install command.

After checking that all code passes the tests, and after removing the vendor directory again, we can run phpcpd again. This time, after refactoring, we will see output that looks like this:

```
$ phpcpd whatstore/
phpcpd 6.0.3 by Sebastian Bergmann.

No clones found.

Time: 00:00.563, Memory: 4.00 MB
```

As you can see, the refactoring has eliminated the duplicated code from the whatstore project. The phpcpd tool is really useful to help us to find duplicated code with TDD.

> **PHP Mess Detector**
>
> There is another interesting CLI tool for PHP named the **PHP Mess Detector (PHPMD)**. This tool looks for several potential problems within PHP source code, such as possible bugs, suboptimal code, overcomplicated expressions, and unused parameters, methods, and properties. It is recommended that you use PHPMD combined with your IDE. You can download PHPMD at https://phpmd.org.

In the next section, we will do some more refactoring, this time to remove unnecessary configurations.

Removing unnecessary configuration

In the previous section, we have learned how to find and remove clones of source code. A clone is an exact copy of the original, like the Jango Fett clones in *Star Wars*. A clone is not a twin brother that is similar but not really identical, like Arnold Schwarzenegger and Danny DeVito in the movie *Twins*. So, phpcpd maybe won't find similar pieces of source code because there is some little difference between them. In our case, there are some similar pieces of source code regarding the view and layout.

As we decided in *Chapter 5*, *Creating the Virtual Store Project*, in the *Preparing the modules for our project* section, the `Application` module will define the application layout. We also said that the last module in the `modules.config.php` array was the module that defines the application layout. But why and how?

You have observed that there are several configuration files in a Laminas MVC application. There are files in the root `config` folder and there is a configuration file for each module. When the application is initialized, all these files are merged in a single configuration array or, better, in a single data map. Each map key is unique, so if the same map key is in different configuration files, one of the values is lost in the merge. The last merged file overrides the previous one, so the last module in the modules list overrides the layout configuration of the others.

But there is no need for overriding values if the same configuration keys are not necessary for every module. This is the case with most of the subkeys of the `view_manager` key in the `module.config.php` file. Only the application module needs all the subkeys because it defines layout and error templates.

So, the `view_manager` key in the `module.config.php` file of the `Application` module must look like this:

```
'view_manager' => [
    'display_not_found_reason' => true,
    'display_exceptions'       => true,
    'doctype'                  => 'HTML5',
    'not_found_template'       => 'error/404',
    'exception_template'       => 'error/index',
    'template_map' => [
        'layout/layout'         => __DIR__ .
            '/../view/layout/layout.phtml',
        'error/404'             => __DIR__ .
            '/../view/error/404.phtml',
        'error/index'           => __DIR__ .
            '/../view/error/index.phtml',
    ],
```

```
        'template_path_stack' => [
            __DIR__ . '/../view',
        ],
```

Observe that we have removed the `application/index` item from `template_map`. This is because this item also is not necessary. In fact, we don't need a page for the `Application/IndexController->indexAction` method. We can remove the `view/application/index` folder and refactor the mentioned method with a redirection:

```
    public function indexAction()
    {
        return $this->redirect()->toRoute('home');
    }
```

So, if somebody types the `/application` path, this person will be redirected to the home page, which is the `Store` module home page.

Well, let us discard the unnecessary view configuration for the other modules, `Inventory` and `Store` (`Generic` has no view). For the `Inventory` module, the `view_manager` key must look like this:

```
    'view_manager' => [
        'template_path_stack' => [
            __DIR__ . '/../view',
        ],
        'strategies' => [
            'ViewJsonStrategy',
        ],
    ],
```

As we learned when we built our API, the `ViewJsonStrategy` value in `strategies` enables controllers to respond with JSON.

For the `Store` module, the `view_manager` key must look like this:

```
    'view_manager' => [
        'template_path_stack' => [
            __DIR__ . '/../view',
        ],
    ],
```

The `template_path_stack` subkey is required because each module has its own view folder.

So, we have removed unnecessary code lines that are not detected by phpcpd because they were not perfect clones. The application keeps working as before, but now it doesn't merge unuseful configuration data.

In the next section, we will do some more refactoring, this time to implement a new feature.

Creating a product search box

The current whatstore home page lists all registered products, but the customer who visits this page would be searching for a specific product. Hence, it is necessary to have a search engine on the page. We could implement this feature because we can see that it is a good feature, but in fact, we will implement it because it is a requirement of our customer. It is the second requirement mentioned in *Chapter 5, Creating the Virtual Store Project* – the home page of the customer area must allow them to search for products by name, completely or partially.

Let us insert an HTML form in an index.phtml file in the module/Store/view/store/index folder. The HTML form will be inside the HTML table that will show the links for the product basket and login, as we can see in the following snippet:

```
<h1>Our products</h1>
<table style="width: 100%">
<tbody>
<tr>
<td><a href="<?=$this->url('store',['controller' =>
    'productbasket'])?>">Product Basket</a></td>
<td>
<form method="post" action="<?=$this->url('store')?>">
<input type="text" name="name"/>
<input type="submit" value="search"/>
</form>
</td>
<td><a href="<?=$this->url('store',['controller' =>
    'customer','action' => $link['action']])?>">
        <?=$link['label']?></a></td>
</tr>
</tbody>
</table>
```

The preceding snippet will render a search box, as we can see in *Figure 12.2*:

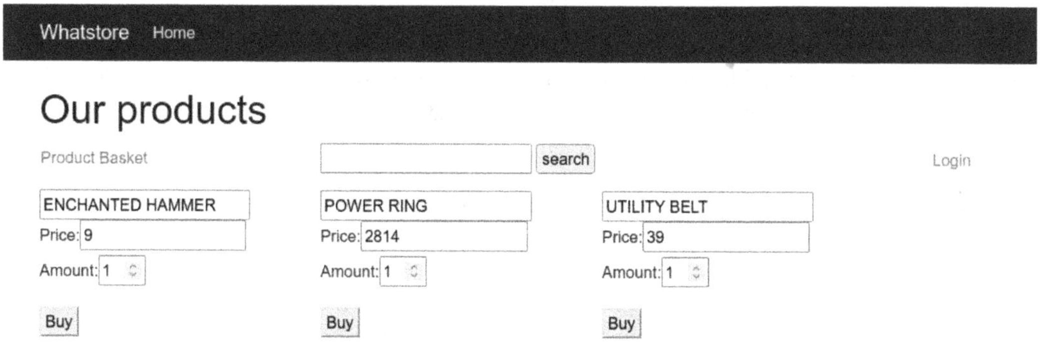

Figure 12.2: The search box on the whatstore home page

The search form will make an HTTP POST request to the `IndexController` default action of the `Store` module. Then, we need to modify the default action, the `indexAction` method, with the following implementation:

```php
public function indexAction()
{
    $name = $this->getRequest()->getPost('name');
    $where = null;
    if (!empty($name)) {
        $where = new Where();
        $name = strtoupper($name);
        $where->like('products.name', "%$name%");
    }
    $products = $this->productTable->getAll($where);
    $form = new ProductForm();
    return new ViewModel([
        'form' => $form,
        'products' => $products
    ]);
}
```

You can see that in the `IndexController->indexAction` method, we have added a check for data provided by an HTTP POST request. If the name field is sent, the data table gateway for products (the `ProductTable` class) receives a `Where` object, configured with a SQL LIKE operator. Laminas'

Where class encapsulates all SQL SELECT operators to filter queries. We have used the `strtoupper` function because product names are stored in uppercase.

Of course, for the `ProductTable->getAll` method to be able to run a SQL SELECT statement, we have to modify the method because the current implementation expects only equality filters. To support the use of LIKE and any other SQL SELECT operator, the `ProductTable->getAll` method must be changed to the following snippet:

```php
public function getAll($where = null): iterable {
    $select = new Select($this->
        tableGateway->getTable());
    $select->join('discounts','products.code_discount
        =discounts.code',['name_discount' => 'name']);
    if (!is_null($where)){
        if (is_array($where)){
            $parsedWhere = [];
            foreach($where as $index => $value){
                $parsedWhere['products.' . $index] =
                    $value;
            }
            $select->where($parsedWhere);
        } else {
            $select->where($where);
        }
    }
    return $this->tableGateway->selectWith($select);
}
```

Remember that the `ProductTable->getAll` method already supports several equality filters joined by an AND operator. Now, this method can use any SQL SELECT operator through a Laminas `Where` class.

With these surgical modifications in three files, the customer is able to search for products by name on the home page.

In the next section, we will learn how to generate API documentation.

Generating API documentation

It is no use having thousands of features if nobody knows about them. Developers need documentation. A developer will probably forget what they did last week, and imagine a new developer entering during the middle of a project. This is an important point.

Another important point is the consumption of our application API by other clients. We talked about this in *Chapter 7, Request Control and Data View*. The creators and maintainers of clients will need good documentation to know how to consume the whatstore services.

In this section, we will create API documentation based on the **OpenAPI Specification**, or **Swagger Specification**. There is an awesome component for generating Swagger documentation from PHP annotations – swagger-php. But swagger-php requires that you write some PHP code to generate Swagger documentation. We will use another component, swagger-docs, that extends swagger-php and requires only the annotations in the classes to be documented.

> Swagger
>
> You can read more about Swagger at `https://swagger.io/resources/open-api`

We will install swagger-docs using Composer, like this:

```
composer require fgsl/swagger-docs
```

You also can use the Eclipse Composer interface, as we learned in *Chapter 3, Using Laminas as a Library with Test-Driven Development*.

We need to add an `@OA\Info` annotation in one single file to define a title and a version for API documentation. We will include this annotation in a new controller class in the `Application` module – `APIController`. The annotation block will be immediately before the class declaration, as we can see in the following complete source code:

APIController.php

```php
<?php
declare(strict_types=1)
namespace Application\Controller;
use Laminas\Mvc\Controller\AbstractActionController;
use Laminas\View\Model\ViewModel;
/**
 * @OA\Info(title="Whatstore API", version="0.1")
```

```
*/
class APIController extends AbstractActionController
{
    public function indexAction()
    {
        $view = new ViewModel();
        $view->setTerminal(true);
        return $view;
    }
}
```

You are probably asking what the setTerminal method is for. This method disables the layout used for this action. We are doing this because the API page won't use the layout.

The APIController->indexAction method requires an index.phtml file. We haven't shown the content of this file here, but it is available in the URL mentioned in the *Technical requirements* section. The model for this API page is based on the template available at https://github.com/fgsl/swagger-docs#model-for-api-page. This template has a placeholder, [WEB ROOT ROUTE], which should be replaced with <?=$this->url('home')?>. This instruction, in its turn, will generate a hyperlink to the web root endpoint.

Next, we will create a specific route for API documentation. In the module.config.php file of the Application module, we will add the following route configuration as one new element for the routes key:

```
'api' => [
    'type'    => Literal::class,
    'options' => [
        'route'    => '/api',
        'defaults' => [
            'controller' => Controller\
                APIController::class,
            'action'     => 'index',
        ],
    ],
],
```

After creating an API route, a controller, and a page without a layout, we can write annotations in the classes. I will show, as an example, the annotation block code for the `ProductAPIController` class of the `Inventory` module. The block will look like this:

```
/**
 * @OA\Post(
 *      path="/inventoryapi/productapi/{key}",
 *      @OA\Parameter(
 *          name="key",
 *          in="path",
 *          description="product code",
 *          required=false
 *      ),
 *      @OA\RequestBody(
 *      @OA\MediaType(mediaType="multipart/form-data",
 *          @OA\Schema(
 *              @OA\
Property(property="code",type="integer"),
 *              @OA\Property(property="name",type="string"),
 *              @OA\Property(property="price",type="float"),
 *              @OA\Property(property="code_
discount",type="integer"),
 *              required={"name","price","code_discount"}
 *          )
 *      )),
 *      @OA\Response(response="200", description="insert
 *      a product"),
 *      @OA\MediaType(mediaType="application/json")
 * )
 */
```

Let us understand each annotation:

- `@OA\Post` informs us that the method after the annotation block makes an HTTP POST request. The argument path is the URL for the web service. The name between braces (`key`) is a variable argument defined by the `@OA\Parameter` annotation.

- `@OA\Parameter` informs us of a parameter that needs to be sent with the request. The `in` argument with the `path` value means that the parameter must be placed in the URL.

- @OA\RequestBody contains the definition for the fields of an HTML form.

- @OA\MediaType within @OA\RequestBody informs us of the value for the enctype attribute of an HTML form.

- @OA\Schema contains a set of form fields. Each field is defined by an @OA\Property annotation.

- @OA\Response informs us of the expected HTTP code and the description of the response.

- @OA\MediaType at the end informs us of the Content-Type header of the HTTP response.

After writing the annotation code block, we can run the swagger-docs command like so:

vendor/bin/fsd module

We pass module as an argument because module is the directory where the classes are. If you wish, you can integrate the fsd command in the Eclipse toolbar using what you learned in *Chapter 2, Setting Up the Environment for Our E-Commerce Application.*

After running the fsd command, we call http://localhost:8000/api, and we will see a page that looks like *Figure 12.3*:

Figure 12.3: The API page with the endpoint to insert products

Observe that the page shows a line with a path to the action that inserts products, preceded by an HTTP method. If we click on the path, the item will be expanded, showing a view that looks like *Figure 12.4*.

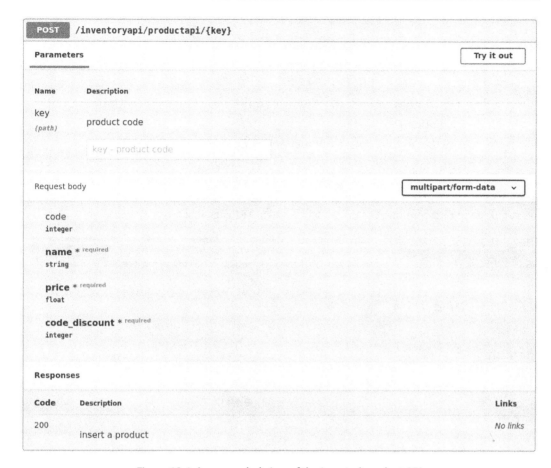

Figure 12.4: An expanded view of the inserted product API

We can make requests from the Swagger view to the whatstore application. You can see that there is a **Try it out** button in the top-right corner. After clicking on this button, the fields of the form become editable, and an **Execute** button appears. Let us see an example of a request on the Swagger page for the recovery of products.

We will add an annotation block code before the declaration of the `ProductAPIController->get` method. The block will look like this:

```
/**
 * @OA\Get(
 *       path="/inventoryapi/productapi/{key}",
 *       @OA\Parameter(
 *              name="key",
```

```
 *           in="path",
 *           description="product code",
 *           required=false
 *      ),
 *      @OA\Response(response="200", description="get a
         product"),
 *      @OA\MediaType(mediaType="application/json")
 * )
 */
```

After running the `fsd` command again, we will have an API page that looks like this:

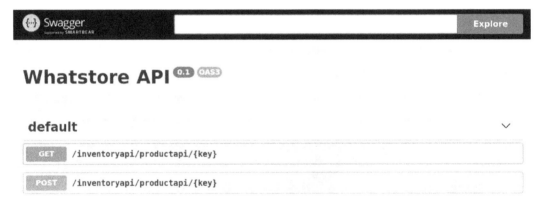

Figure 12.5: An API page with endpoints for the GET and POST methods

If we click on the item for HTTP GET request of products, the page will expand the item, showing a form, as we saw in *Figure 12.4*. When we click on the **Try it out** button, two buttons appear, **Execute** and **Clear**, as we can see in *Figure 12.6*.

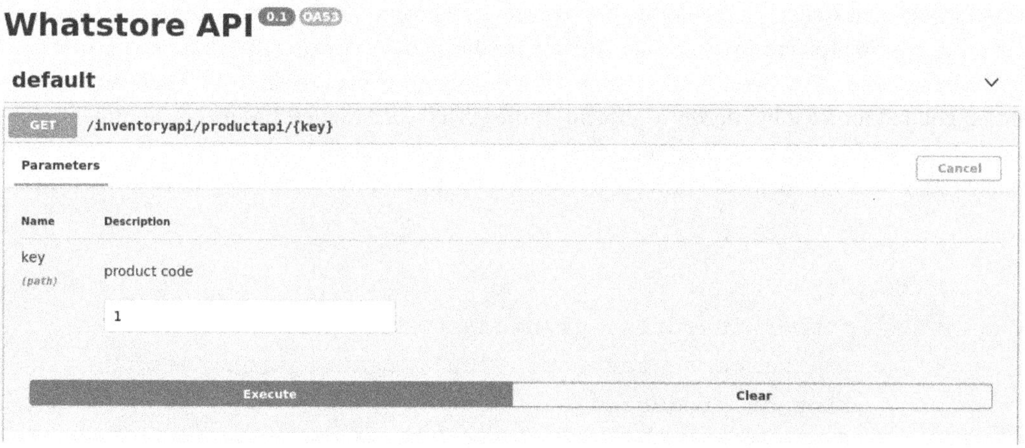

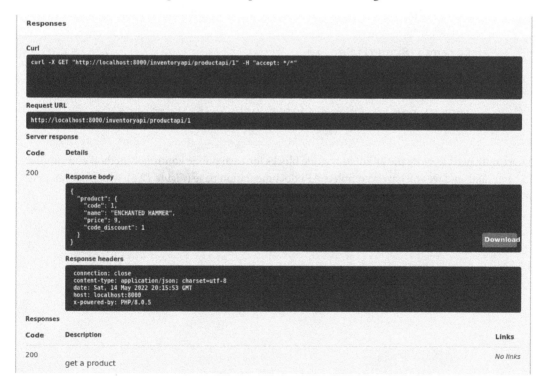

Figure 12.6: A form to make an HTTP GET request

After filling the product code field with a valid value and clicking on the **Execute** button, the page will show a section with a response to the request, as we can see in *Figure 12.7*.

Figure 12.7: The output response section on the Swagger page

Currently, note that we have a method to get one product; however, we don't have a method to get a list of products. No problem. We can add a method for this to the `ProductAPIController` class. The `AbstractRestfulController` superclass distinguishes between an HTTP GET request for one element and for a list by the key argument. If the HTTP GET request has no key argument, the `AbstractRestfulController` superclass executes the `getList` method. Then, we implement this method, which will look like this:

```
/**
 * @OA\Get(
 *      path="/inventoryapi/productapi",
 *      @OA\Response(response="200", description="get
 *      list of products"),
 *      @OA\MediaType(mediaType="application/json")
 * )
 */
public function getList()
{
    $resultSet = $this->productTable->getAll();
    $products = [];
    foreach($resultSet as $result){
        $products[] = $result->toArray();
    }
    return new JsonModel(['products' => $products]);
}
```

Now, we only need to write the annotation code blocks for each public controller method and run the `fsd` command, and we will have a complete API documentation page (*Figure 12.8*), which additionally allows us to test the requests for the whatstore application.

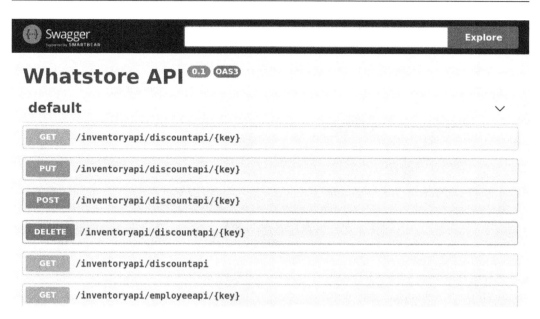

Figure 12.8: The top of a complete API documentation page generated with Swagger

In the next and final section, we will review the authentication implementations of whatstore and refactor them to avoid security issues.

Isolating identity management for each module

In *Chapter 9, Event-Driven Authentication*, we learned how to implement an authentication control for employees in the `Inventory` module. In *Chapter 11, Implementing a Product Basket*, we learned how to implement an authentication control for customers in the `Store` module. So, the whatstore application has two authentication controls, both of them using the `laminas-authenticate` component.

The authentication control is based on events. Specifically, it is based on the Laminas `EVENT_ROUTE` event. The identity check is done through listener classes, specifically `InventoryAuthenticationListener` in the `Inventory` module and `StoreAuthenticationListener` in the `Store` module. Both of these two classes check identity with the same lines of code:

```
$authenticationService = new AuthenticationService();
if (!$authenticationService->hasIdentity()){
```

In fact, `InventoryAuthenticationListener` does not have these lines so instead calls the `IdentityManager->hasIdentity` method, which does have them. However, ultimately, both of the listener classes use the same lines of code to check the authentication state.

Looking at this code, we can see that the `AuthenticationService->hasIdentity` method returns a Boolean value. Maybe you didn't need to worry about this before, but think about it now – how does `AuthenticationService` know whether there is an authenticated identity?

Here, we have an example of one of the advantages of free and open source software – we can see how things work. If we find the source code of the `AuthenticationService->hasIdentity` method, we will see the following lines:

```
public function hasIdentity()
    {
        return ! $this->getStorage()->isEmpty();
    }
```

Note that `AuthenticationService` answers that there is an authenticated identity if its storage is not empty. However, the storage content is not tested. This is not a problem if there is only one authentication control, but it can cause unexpected behavior for systems with more than one authentication control, such as whatstore.

What is this unexpected behavior? Imagine that a customer is authenticated in the `Store` module. This customer, Mr. Bad Guy, specifically, is someone with bad intentions. He tries to call some endpoints to discover some interesting URLs until he discovers that there is an `/inventory` route. The expected behavior is that the `/inventory` route shows a login page. However, Mr. Bad Guy, who has been authenticated in the `Store` module, goes directly to the menu page of the `Inventory` module. It does not mean that Mr. Bad Guy will get to access the `Inventory` registrations, because there are authorization controls too, but this direct access to the menu page is clearly a security failure.

What is the cause of the problem? It's the use of the `AuthenticationService->hasIdentity` method for two authentication controls in the same application. The `laminas-authentication` documentation presents samples for only one authentication control. This suggests that the `AuthenticationService->hasIdentity` method is usually used for only one authentication control. In fact, you can use the `laminas-authentication` component for as many authentication controls as you want. It is only necessary to isolate each authentication context. This means avoiding the direct use of the `AuthenticationService->hasIdentity` method to check whether authentication was done successfully. We will do this next.

Modifying the class for identity management

We have already created a generic identity management class (`IdentityManager`, in the `Generic` module) and extended it for each module. Now, we will modify this class to manage identities based on a context – more specifically, the namespace of the identity manager class, which will be different for each module. Let us modify the `Generic\Model\IdentityManager` class in three steps.

First, let us add a getNamespace method to the class:

```
protected function getNamespace(): string
{
    $fqcn = explode('\\', static::class);
    return $fqcn[0];
}
```

This method will return the namespace of the called class. This means that it is not exactly a Generic namespace. If the method is called by a subclass, the considered namespace will be the namespace of the subclass. For example, if the getNamespace method is called by Inventory\Model\ IdentityManager, it will return Inventory.

The second step is modifying the Generic\Model\IdentityManager->__construct method:

```
public function __construct(AdapterInterface $adapter,
array $encryptionMethod)
{
    $this->adapter = $adapter;
    $this->encryptionMethod = $encryptionMethod;
    $this->auth = new AuthenticationService();
    $storage = new Session($this->getNamespace());
    $this->auth->setStorage($storage);
}
```

Note that we have added two new lines to the constructor method. We instantiate the Laminas\ Authentication\Storage\Session class, defining as a namespace the value of the namespace returned by the getNamespace method. So, we will have a namespace inside the session, avoiding possible session key overriding for different IdentityManager instances.

The third step is modifying the Generic\Model\IdentityManager->hasIdentity method:

```
public function hasIdentity(): bool
{
    if ($this->auth->hasIdentity()) {
        $storage = $this->auth->getStorage();
        error_log('namespace: ' . $storage-
            >getNamespace());
        return ($storage->getNamespace() === $this-
            >getNamespace());
    }
}
```

```
        return false;
    }
```

Now, we are not only checking whether there is an authenticated session but also whether the storage session namespace is the same as the namespace of the IdentityManager instance. In the next section, we will use this new IdentityManager class to refactor the StoreAuthenticationListener and IndexController classes of the Store module.

Refactoring listener and controller classes

In this section, we will refactor some lines of the Store\Listener\StoreAuthenticationListener and Store\Controller\IndexController classes. In the StoreAuthenticationListener class, we will first locate these lines:

```
$authenticationService = new AuthenticationService();
if (!$authenticationService->hasIdentity()){
```

Replace them with the following lines:

```
$identityManager = $event->getApplication()->
    getServiceManager()->get(IdentityManager::class);
if (!$identityManager->hasIdentity()){
```

Remember that IdentityManager, in this case, is specifically Store\Model\IdentityManager.

In Store\Controller\IndexController, we will add a new attribute and modify the constructor method to use this attribute:

```
    private IdentityManager $identityManager;
    public function __construct(ProductTable $productTable,
        IdentityManager $identityManager)       {
        $this->productTable = $productTable;
        $this->identityManager = $identityManager;
    }
```

In the Store\Controller\IndexController->indexAction method, we will add a key into the array injected into the ViewModel instance:

```
return new ViewModel([
        "form" => $form,
        "products" => $products,
        "authenticated" => $this->identityManager-
```

```
            >hasIdentity()
]);
```

Of course, if we have modified the controller constructor method, we have to adjust its factory method in `Store\Controller\IndexControllerFactory`:

```
    public function __invoke(ContainerInterface $container,
        $requestedName, array $options = null)    {
        $identityManager = $container->get
            (IdentityManager::class);
        return new CustomerController($identityManager);
    }
```

Now, we will locate these lines of the `view\store\index\index.phtml` file:

```
$authenticationService = new AuthenticationService();
$link = ["action" => "index", "label" => "Login"];
if ($authenticationService->hasIdentity()){
```

Replace them with the following lines:

```
$link = ["action" => "index", "label" => "Login"];
if ($this->authenticated){
```

So, we avoid the coupling of the view script with an authentication class, limiting the view script to use variables, which is more appropriate. The good practice for view scripts is only to use variables sent by a controller class.

So, we have finished the needed changes to isolate the two authentication controls of the whatstore application. We don't need to worry about Mr. Bad Guy anymore.

In this section, we have isolated identity management to avoid a user of a module having privileged and disallowed access to another module. To implement this, we modified the generic class for identity management, `Generic\Model\IdentityManager`. Finally, we refactored the `StoreAuthenticationListener` and `CustomerController` classes to use the inherited `Store\Model\IdentityManager` class.

Summary

In this chapter, we started by discussing some agile development concepts, which are very important for software maintenance. After these concepts, we learned how to easily detect duplicated code in PHP programs using the free and open source `phpcpd` tool. After detecting blocks of duplicated code to filter and validate in model classes, we showed how to unify these blocks in an extension of the `Laminas\InputFilter\InputFilter` class.

Next, we learned how to create a product search box. For this, we had to modify the `ProductTable` class to execute SQL `SELECT` queries with the `LIKE` operator.

Then, we learned how to generate API documentation using the free and open source swagger-docs tool. When we demonstrated how to use the OpenAPI Specification to generate rich HTML pages with REST endpoints, we also learned how to implement REST requests to list collections.

Finally, we learned how to isolate identity management to avoid security issues when we have more than one authentication control.

In the next chapter, we will introduce some tips and tricks that help devselopers to solve some problems that can eventually appear in the construction of an object-oriented PHP web application.

13
Tips and Tricks

This chapter provides some tips and tricks that will help you as a developer to solve some problems that may eventually appear in the construction of a web application.

We will first start by learning what the required methods for creating mapped models, that is, classes that are connected to database tables, are.

After that, we will learn how to customize filters and validators to implement specific business rules in the model layer.

Then, we will understand the Laminas view layer better, learning how to produce JSON responses, manage file uploads, and change the application layout.

Finally, we will be mastering the Laminas controller layer, by learning how to detect AJAX requests, how to exchange messages between the controller and view layers, and what the different ways of ending a controller action are.

By the end of this chapter, you will be able to meet eventual demands on web applications using the Laminas framework.

In this chapter, we'll be covering the following topics:

- Creating mapped models
- Customizing filters and validators
- Understanding the Laminas view layer
- Mastering the controller layer

Technical requirements

All the code related to this chapter can be found at two repositories:

- `https://github.com/PacktPublishing/PHP-Web-Development-with-Laminas/tree/main/chapter13/tipsandtricks`

- `https://github.com/PacktPublishing/PHP-Web-Development-with-Laminas/tree/main/chapter13/whatstore`

The first URL contains a Laminas application named `tipsandtricks`. All of the code samples in this chapter can be found in `tipsandtricks`. You can run this application with PHP as a web server, as we did in *Chapter 2, Setting Up the Environment for Our E-Commerce Application*, with the **myfirstlaminas** application.

The second URL is for our e-commerce application, `whatstore`, with a little modification related to the use of filters. This will be explained in the *Creating a customized filter* section.

Creating mapped models

This is a useful tip for avoiding headaches when you are creating models mapped to tables with `laminas-db`. These models need to implement two methods, that is, `exchangeArray` and `getArrayCopy`:

```
public function exchangeArray($data)
```

The `exchangeArray` method is used by the `Laminas\Db\ResultSet` class to populate a model with data from a table record.

```
public function getArrayCopy()
```

The `getArrayCopy` method is used by the `Laminas\Db\ResultSet` class to get data from a model. It is also used by `Laminas\Form\Form` to bind model attribute values to form field values.

In fact, the `Laminas\Db\ResultSet` class also accepts a `toArray` method, replacing the `getArrayCopy` method. I suggest implementing the `toArray` method in models to extract the data to be persisted with the `TableGateway` class. Sometimes, the model data to be sent to a form object is different from the model data to be sent to a data mapper. You can show, in an HTML form, data that won't be saved in a database table, because an HTML form can show data from two joined tables and one of these tables is read-only, not written.

This was a short tip. It was a good start. In the next section, still talking about the model layer, we will learn how to customize filters and validators. It won't be as short, but it will be more exciting.

Customizing filters and validators

In this section, we will learn how to customize filters and validators and what the required methods for mapped models are.

On the home page of the tipsandtricks application, you will see a frame, similar to that in *Figure 13.1*, on the left-hand side, with samples of filter and validator customization for the model layer:

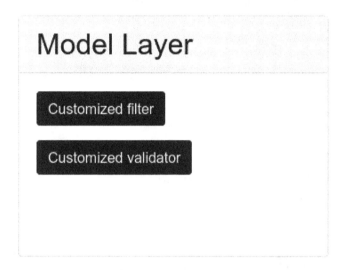

Figure 13.1 – Access to customized filter and validator samples

We will learn how the buttons of this frame work and understand how filters and validators are created with the Laminas framework.

In *Chapter 8, Creating Forms and Implementing Filters and Validators*, we learned how to use some filters and validators available in the `laminas-filter` and `laminas-validator` components. However, we are not limited to the implemented Laminas filters and validators. Both of these components have interfaces that allow us to implement customized filters and validators.

Let us test the customized filter first.

Creating a customized filter

Click on the **Customized filter** button (shown in *Figure 13.1*) and you will be taken to the following page:

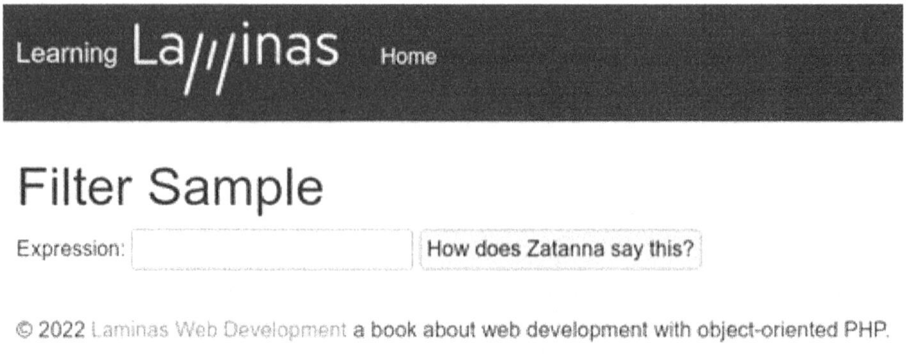

Figure 13.2 – Form to test the customized filter

The form shown in *Figure 13.2* serves to apply a filter that reverses the typed expression. As an example, you can type open sesame in the text box and then click on the **How does Zatanna say this?** button. In addition to finding out that Zatanna has some pronunciation problems, you will see the page reload with the following content:

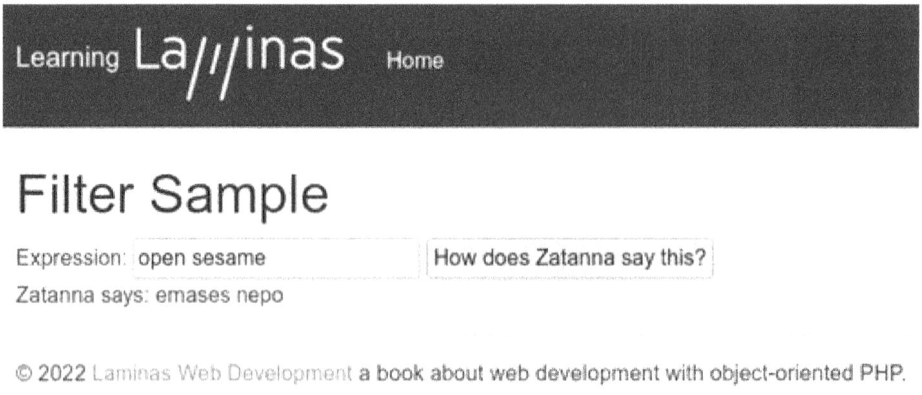

Figure 13.3 – Form page with the applied filter

The form in *Figure 13.3* calls the IndexController->filterAction method, implemented as follows:

```
public function filterAction()
{
    $expression = $this->getRequest()->getPost
```

```
            ('expression');
        if (empty($expression)){
            $response = '';
        } else {
            $filter = new Zatanna();
            $response = 'Zatanna says: ' . $filter->filter
                ($expression);
        }
        return new ViewModel([
            'expression' => $expression,
            'response' => $response
        ]);
    }
```

Observe that the `filterAction` method creates an instance of a `Zatanna` class if the form has sent a value. The `Zatanna` class has a `filter` method, which reverses the given value. The `Zatanna` class implements the `Laminas\Filter\FilterInterface` interface, which is common for every Laminas filter class:

Zattana.php

```php
<?php
namespace Application\Filter;
use Laminas\Filter\FilterInterface;
class Zatanna implements FilterInterface
{
    public function filter($value)
    {
        return strtolower(strrev($value));
    }
}
```

As you can see, a filter class is a class that must implement a `filter` method, which receives an untyped value and returns something. So, you can create your own filter classes according to your needs and use them easily.

Filter for encrypting passwords

As you should remember, we created methods for encrypting a password in the `Customer` and `Employee` classes. The encrypting implementation was the same for the two model classes. In the whatstore version of this chapter, you will find an example of a customized filter applied to extract the duplicated encrypting implementation for a single class. Search for the `Encrypt` class in the `Generic` module.

In the next section, we will learn how to customize validators.

Creating a customized validator

Go back to the tripsandtricks home page and click on the **Customized validator** button; you will be taken to the following page:

Figure 13.4 – Form to test the customized validator

The form shown in *Figure 13.4* serves to apply a validator that checks whether the typed word or phrase is a palindrome. As an example, you can type `Namor sees Roman` in the text box and then click on the **Is a palindrome?** button. Then, you will see the page reload with the following content:

Figure 13.5 – Form page with the applied validator

The form shown in *Figure 13.5* calls the `IndexController->validatorAction` method, implemented as follows:

```php
public function validatorAction()
{
    $expression = $this->getRequest()->getPost
        ('expression');
    if (empty($expression)){
        $response = '';
    } else {
        $validator = new Palindrome();
        $response = 'It is not a palindrome!';
        if ($validator->isValid($expression)){
            $response = 'It is a palindrome!';
        }
    }
    return new ViewModel([
        'expression' => $expression,
        'response' => $response
    ]);
}
```

Observe that the `validatorAction` method creates an instance of the `Palindrome` class if the form has sent a value. The `Palindrome` class has an `isValid` method, which returns `true` if the given value is a palindrome. The `Palindrome` class implements the `Laminas\Validator\ValidatorInterface` interface, which is common for every Laminas validator class:

Palindrome.php

```php
<?php
namespace Application\Validator;
use Laminas\Validator\ValidatorInterface;
class Palindrome implements ValidatorInterface
{
    private array $messages = [];
    public function isValid($value)
    {
        $value = strtolower($value);
```

```php
        $value = str_replace(' ', '', $value);
        $valueLength = strlen($value);
        $repeated = (int)($valueLength/2);
        $n = $valueLength-1;
        for($i=0;$i<$repeated;$i++){
            $palindrome = ($value[$i] == $value[$n]);
            if (!$palindrome){
                $this->messages[] = "{$value[$i]} is
                    different of {$value[$n]}";
                return false;
            }
            $n--;
        }
        return true;
    }
    public function getMessages()
    {
        return $this->messages;
    }
}
```

As you can see, a validator class is a class that implements an `isValid` method, which receives an untyped value and returns `true` or `false`, and a `getMessages` method, which returns an array of error messages. So, you can create your own validator classes as per your needs and use them easily.

Next, we will understand the Laminas view layer better.

Understanding the Laminas view layer

In this section, we will learn how to produce JSON responses in controller actions and how to upload files using `laminas-form`.

At the center of the tipsandtricks home page, you can see a frame that looks like in *Figure 13.6*, with samples of the features mentioned in the previous section for the view layer:

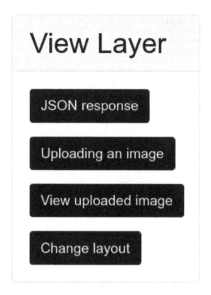

Figure 13.6 – Access to view samples

We will use the actions related to this frame in the following subsection.

Producing JSON responses with laminas-view

To create HTTP responses with data in JSON format, you need to do two things:

1. Use the Laminas\View\Model\Json class as the return type of the controller action method.
2. Add the following element to the view_manager array key of the module.config.php file:

```
'strategies' => [
        'ViewJsonStrategy',
    ],
```

Let us see an example of a JSON response. Click on the **JSON response** button in the center frame. You will see the following output:

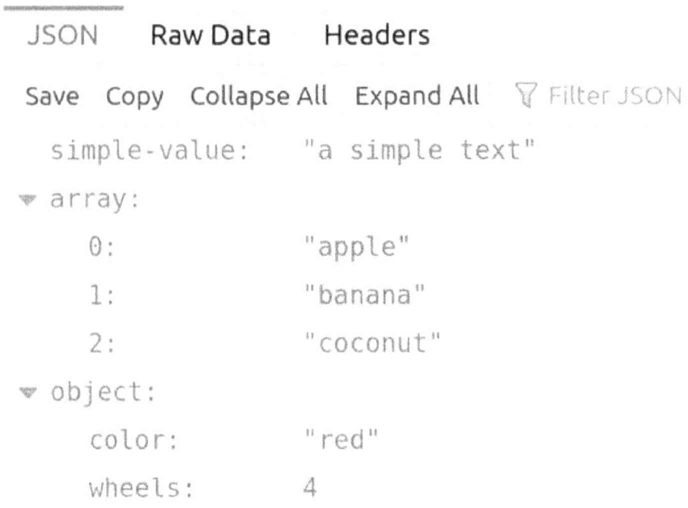

JSON   Raw Data   Headers

Save   Copy   Collapse All   Expand All   ▽ Filter JSON

   simple-value: "a simple text"

▼ array:
     0:         "apple"

     1:         "banana"

     2:         "coconut"

▼ object:
    color: "red"

    wheels: 4

Figure 13.7 – JSON response generated by JsonModel

This response is generated by the `IndexController->showJsonAction` method, as follows:

```php
public function showJsonAction()
{
    $simpleValue = 'a simple text';
    $list = ['apple','banana', 'coconut'];
    $car = new \stdClass();
    $car->color = 'red';
    $car->wheels = 4;
    return new JsonModel([
        'simple-value' => $simpleValue,
        'array' => $list,
        'object' => $car
    ]);
}
```

Since `ViewJsonStrategy` is enabled, the values injected into the `JsonModel` constructor are converted into JSON format and the content type of the HTTP response is configured for JSON.

Do you have déjà vu? Yes! It is because we already used `ViewJsonStrategy` in *Chapter 7, Request Control and Data View*. But we were worried that you might have forgotten this. In an API-oriented world, it is good to know how to generate JSON responses.

Next, let us learn how to handle a file upload.

Uploading files with laminas-form

You've probably realized that something is missing in the product registration of your whatstore application: the product images. Images can be stored in database tables, using text or image-specific fields. Images can also be stored in the filesystem, so the database can retain only the paths for image files. But no matter how the images are stored or how many images are associated with a product, the upload process for a file is the same.

If you click on the **Uploading an image** button, you will be taken to the following page:

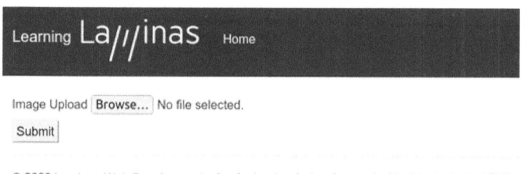

Figure 13.8 – Form to upload an image

The form for uploading is created by the UploadForm class, which extends Laminas\Form\Form:

UploadForm.php

```php
<?php
namespace Application\Form;
use Laminas\Form\Element;
use Laminas\Form\Form;
class UploadForm extends Form
{
    public function __construct($name = null, $options = [])
    {
        parent::__construct($name, $options);
        $this->addElements();
    }
    public function addElements()
    {
```

```
        // File Input
        $file = new Element\File('image-file');
        $file->setLabel('Image Upload');
        $file->setAttribute('id', 'image-file');
        $this->add($file);
    }
}
```

The **Browse...** button is rendered from the Laminas\Form\Element\File class. When you click on the **Browse...** button, the browser will open the file manager window and you can select a file. The tipsandtricks application has a sample royalty-free image file named pets.jpg in the folder data. After selecting the image file and clicking on the **Submit** button, the page is reloaded with the addition of the uploaded data:

```
Array
(
    [image-file] => Array
        (
            [name] => pets.jpg
            [type] => image/jpeg
            [tmp_name] => /opt/lampp/temp/phpQb1F1p
            [error] => 0
            [size] => 226510
        )
)
```

Observe that we have, as main data, the former name of the uploaded file, the type of file, and the temporary directory in the server where the file was stored. This data is sent by the IndexController->uploadFormAction method, as follows:

```
    public function uploadFormAction()
    {
        $form = new UploadForm('upload-form');
        $form->setAttribute('action',$this->url()->
            fromRoute('application',
            ['action' => 'upload-form']
        ));
        $data = [];
```

```
        $request = $this->getRequest();
        if ($request->isPost()) {
            // Make certain to merge the $_FILES info!
            $post = array_merge_recursive(
                $request->getPost()->toArray(),
                $request->getFiles()->toArray()
            );
            $form->setData($post);
            if ($form->isValid()) {
                $data = $form->getData();
                $content = file_get_contents($data
                    ['image-file']['tmp_name']);
                file_put_contents(self::TARGET_DIR .
                    'image.jpg', $content);
            }
        }
        return new ViewModel([
            'form' => $form,
            'data' => $data
        ]);
    }
```

If the data form is valid, we copy the image file from the temporary directory to the `public/img` directory of the application.

The reloaded page also has a **See the uploaded image** link. Click on this link and you will be taken to the following page:

Figure 13.9 – Uploaded image

As you can see, it is very easy to upload files with `Laminas\Form\Element\File`.

> **Security recommendations**
>
> Use the `Laminas\Filter\File\RenameUpload` component to change the target directory for uploaded files. This prevents an attacker from trying to call a file because the default upload directory is known. It is also good to use the `Laminas\Validator\File\MimeType` component to check whether the uploaded file is really of the expected type.

In the next section, we will see how it is also easy to change the application layout.

Changing the application layout

At the bottom of *Figure 13.6*, you can see a **Change layout** button. If you click on this button, you will be taken to the following page:

Other layout   Home

Change layout sample

Look at the page header. We are using other layout.

© 2022 Laminas Web Development a book about web development with object-oriented PHP.

Figure 13.10 – Alternative layout

As you can observe, the header color and content are different, as well as the body background color. This layout change is made by the `IndexController->changeLayoutAction` method, which calls the `setTemplate` method of the layout controller plugin:

```
public function changeLayoutAction()
{
    $this->layout()->setTemplate('/layout/otherlayout');
    return new ViewModel();
}
```

The `setTemplate` method changes the layout only for the current action. If you click on the home page link and navigate to other pages, you will see that the default layout is still used.

The `setTemplate` method searches for the layout file from the `view` folder. The file extension is `.phtml`, but you don't need to put the extension when calling the method.

You can also disable the layout for a controller action. We saw how to do this in *Chapter 12, Reviewing and Improving Our App*, when we created the API page. The layout is disabled from the `ViewModel` class, with the `setTerminal` method receiving `true`, as follows:

```
$view = new ViewModel();
$view->setTerminal(true);
```

So, you have complete control over the application layout.

In the next section, we will learn some useful things about Laminas for the controller layer.

Mastering the controller layer

In this section, we will learn how controller classes detect AJAX requests, how you can exchange messages between the controller and view layers, and how you can finish a controller action.

On the home page of the tipsandtricks application, you will see the frame shown in *Figure 13.11* with samples of the mentioned features on the right side under the **Controller Layer** heading:

Figure 13.11 – Access to controller layer samples

We will use the actions related to this frame in the following subsection.

Detecting AJAX requests

The **Asynchronous request** and **Synchronous request** buttons make HTTP requests to the same controller action, the IndexController->ajaxAction method:

```
public function ajaxAction()
{
    if ($this->getRequest()->isXmlHttpRequest()){
        return new JsonModel();
    }
    return new ViewModel();
}
```

The `Request->isXmlHttpRequest` method returns `true` if the request was made with XMLHttpRequest. This object makes asynchronous requests. As XMLHttpRequest is handled by JavaScript on the client side, and XML was used a lot to exchange data between the client and server sides, the use of XMLHttpRequest received a special name: **Asynchronous JavaScript and XML (AJAX)**. Although XML use is increasingly rare nowadays because of the lightness of JSON, the acronym AJAX remains in use.

> **XMLHttpRequest specification**
>
> XMLHttpRequest is an open web standard documented at `https://xhr.spec.whatwg.org`.

If you click on the **Asynchronous request** button, you will see an alert window, as shown in *Figure 13.12*. It shows that the page has not been reloaded:

Figure 13.12 – Response to asynchronous requests

If you click on the **Synchronous request** button, you will be taken to the following page:

Figure 13.13 – Response to synchronous requests

You can see that the page has been reloaded. Observe that the same controller action can handle synchronous and asynchronous requests, generating different responses as per the situation.

In the next subsection, we will see a way to exchange messages between a controller and a view script without using the `ViewModel` class.

Exchanging messages between the controller and view layers

If you click on the **Flash messenger** button, you will be taken to the following page:

Figure 13.14 – Message stored by the FlashMessenger plugin

The **Flash messenger** button calls the IndexController->flashMessengerAction method like so:

```
public function flashMessengerAction()
{
    $this->flashMessenger()->addMessage('You called
        this action at ' . date('H:i:s'));
    $messages = $this->flashMessenger()->getMessages();
    return new ViewModel(['messages' => $messages]);
}
```

The FlashMessenger class used in the controller stores messages in the session container. The messages are stored in a message stack, which can be accessed by the getMessages method. In short, the FlashMessenger class is a message manager. The normal way to send the messages stored in an instance of FlashMessenger to the view script would be to inject the output of the getMessages method in the ViewModel constructor, as we can see in the previous code snippet. But as we can see in the following code snippet, this is not necessary:

flash-messenger.phtml

```
<h1>Messages</h1>
<?=$this->flashMessenger()->render()?>
```

Observe that the view script related to the `flashMessengerAction` method does not use the `$messages` variable injected in `ViewModel`. Instead of this, the view script uses a view helper of `FlashMessenger`, which has a `render` method. The `render` method avoids iterating over a message array, outputting the content as simple text.

`FlashMessenger` is a controller plugin, and it is not a part of `laminas-mvc`. You have to install it via Composer separately.

There are two ways to use a not-embedded controller plugin. The easiest is to add the plugin as a module in the `modules.config.php` file. You must add the **Fully Qualified Class Name (FQCN)**:

```
'Laminas\Mvc\Plugin\FlashMessenger'
```

The alternative and more verbose way is to configure the plugin in the `module.config.php` file of the module where it will be used. You need to create a `controller_plugins` key, with `factories` and `aliases` subkeys. The following code snippet shows how the `FlashMessenger` plugin could be configured in the tripsandtricks application:

```
'controller_plugins' => [
    'factories' => [
        FlashMessenger::class => InvokableFactory::
            class
    ],
    'aliases' => [
        'flashMessenger' => FlashMessenger::class
    ]
],
```

While the `factories` key defines how to create the `FlashMessenger` instance, the `aliases` key defines the name that you will use to call the plugin in the controller actions.

Embedded controller plugins

There are some embedded controller plugins that don't need configuration to use. We already used some of them, such as `$this->params()` to get parameters from the URL and `$this->layout()` to change the layout configuration. There is also `$this->url()->fromRoute()`, which allows generating the URL from routes.

This is the configuration for the controller plugin. To use the `FlashMessenger` view helper, if you didn't configure the component as a module, you need to add a `view_helpers` key to the `modules.config.php` file, as follows:

```
'view_helpers' => [
    'invokables' => [
        'flashMessenger' => \Laminas\Mvc\Plugin\
        FlashMessenger\View\Helper\FlashMessenger::class
    ]
]
```

The controller plugin configuration should be new to you, but the view helper configuration is not. In fact, you already coded this configuration in *Chapter 8*, *Creating Forms and Implementing Filters and Validators*, to enable form view helpers.

Next, we will see the different ways to end controller actions.

Finishing controller actions

In most cases, a controller action can terminate when one of the following objects is returned:

- A `ViewModel` instance, if it is a response for a synchronous request, which expects an HTML page.

- A `JSONModel` instance, if it is a response for an asynchronous request, which expects JSON data.

- A call to the `$this->redirect()->toRoute()` method. This method expects two arguments; the first is the name of the route and the second is an array with route parameters. This method causes an HTTP redirecting request (HTTP status 302).

There are other `Laminas\View\Model` classes: `ConsoleModel`, for generating output for terminals, and `FeedModel`, for generating output for RSS consumers. You can also extend `Laminas\View\Model ModelInterface` and create specific outputs.

Summary

In this chapter, we learned some useful tips and tricks relating to Laminas in the three layers of an MVC application.

We first learned how to customize filters and validators for model layers using the Laminas interfaces. We learned about the required methods for using mapped models or models integrated with forms.

After that, we learned, or better, we remembered, how to produce JSON responses for the view layer. We also learned how to upload files using `laminas-form` and how to change the layout configuration.

At the end, we learned how to detect AJAX requests in controllers and how to manage messages with `laminas-mvc-plugin-flashmessenger`. Finally, we learned the possible ways to end a controller action.

In the next chapter, we will cover a final few things. Yes, it is a pity, but this book is coming to an end...

14

Last Considerations

This chapter marks the conclusion of the book, showing the possibilities going forward with other components not used in sample applications.

In this chapter, we will get a general view of Laminas components that were not used in get sample applications in earlier chapters of the book. We will learn about a component for creating distributed and decentralized applications, as well as learning about a component for creating APIs. Lastly, we will learn where we can get help with Laminas components and how to become a part of the Laminas community and improve the Laminas framework.

In this chapter, we will cover the following topics:

- Other Laminas components
- Microservice-oriented development
- Building APIs
- Resources of community

Other Laminas components

In this section, we will get an overview of some Laminas components that were not used in the projects of this book. You can read the documentation about all the Laminas components at `https://docs.laminas.dev/components`.

Cache

Let's use travel as an analogy. We spend time and money traveling. Since time is money, we spend a lot of money traveling. In addition, the farther we go, the more money we spend. In other words, the cheapest place to go is where you are already. Staying in the same place is the cheapest type of travel. This is the idea of the cache: to keep data in the nearest place possible to reduce the time and cost of transport. The `laminas-cache` component allows us to cache data using several storage adapters, which enable your application to save and read data from filesystem, Memcached, Redis, or MongoDB databases, among others.

CLI

Although we have used Laminas to develop web applications, the framework is not limited to this purpose. In fact, the PHP manual says that PHP can do anything. With Laminas, we can do anything, but with more control.

> **What can PHP do?**
> You can find a description of the possible uses of PHP at https://www.php.net/manual/en/intro-whatcando.php.

Believe it or not, you can create command-line PHP programs with Laminas. To do this, you should use the laminas-cli component.

Crypt

In our whatstore project, we said that it is important to encrypt passwords. But you shouldn't use the algorithms that were used in the examples in this book. It is recommended to use strong cryptography and password hashing to protect and authenticate sensitive data. The laminas-crypt component offers an easy and secure way to implement cryptography in PHP applications. This component was developed by several great applied cryptography specialists, such as Enrico Zimuel, a core member of the PHP Framework Interoperability Group.

CSRF

There is actually no component called laminas-crsf. However, there are two classes that handle CSRF. **CRSF** stands for **cross-site request forgery**, an attack where someone tries to send malicious data through falsified forms. The laminas-form component provides Element\Csrf to add a hash element to a form with timeout configuration. To validate the hash element value, you can use the Laminas\Validator\Csrf class.

DOM

DOM stands for **document object model**. This is a web standard, documented at https://www.w3.org/DOM/DOMTR. According to this standard, the HTML tags can be handled as an object hierarchy. In addition, it is possible to search for and select objects using the XPath language and a CSS selector. The laminas-dom component enables you to find elements in XML and HTML documents using these two approaches. It is very useful to automate the reading of documents in these formats in order to import data or search for data on websites through the implementation of web crawlers.

> **XPath and CSS selectors**
>
> The root documentation for the XPath language is available at `https://www.w3.org/TR/1999/REC-xpath-19991116`. The root documentation for CSS selectors is available at `https://www.w3.org/TR/2018/REC-selectors-3-20181106`.

> **The laminas-dom development state**
>
> The `laminas-dom` component is considered feature-complete and is now in **security-only** maintenance mode, following a decision by the **Laminas Technical Steering Commitee**. It does not mean that the package has been abandoned, but it means that only security updates will be incorporated. However, this committee suggests `symfony/dom-crawler` as an alternative component with active development (that is, feature-incomplete).

> **What to do with abandoned packages**
>
> Although `laminas-dom` has not been abandoned, it is useful to know what to do if a component is no longer under development. If the maintainers of a component are worried about their users, you can find information about alternative components on the GitHub home page of the component or on its Packagist page.

Escaper

According to the **Open Web Application Security Project** (**OWASP**), injection was one of the top 10 web application security risks in 2021. One of the forms of injection is **Cross-Site Scripting** (**CSS**). To prevent this type of injection attack in your PHP application, you can use the `laminas-escaper` component. This component neutralizes injected code with bad intentions for output data, no matter whether the code is written in HTML, CSS, or JavaScript.

> **Top 10 OWASP**
>
> OWASP is a foundation that aims to improve software security. You can find a complete rundown of security risks, example attack scenarios, and how to prevent them at `https://owasp.org/www-project-top-ten`.

Feed

If you need to consume RSS or Atom feeds, you can use the `laminas-feed` component. This component enables you to handle feed responses as object collections.

> **RSS reference**
>
> You can find documentation about RSS at `https://validator.w3.org/feed/docs/rss2.html`.

Hydrator

In an object-oriented application, there are some frequent operations, such as the following:

- Hydrating objects (populating them with a dataset)
- Extracting data from objects

The `laminas-hydrator` component provides hydrating and extracting functionalities. This component requires the following from the classes to be hydrated and extracted: they must implement the `exchangeArray` and `getArrayCopy` methods, the same required methods for mapped models, as we saw in *Chapter 13, Tips and Tricks*.

JSON-RPC server

JSON is the de facto standard for REST APIs. But these aren't my words; they're the words of Balaji Varanasi and Sudha Belida in their book, *Spring REST*, from *Apress*, published in 2015. They are right. Nobody will expose API data in a format other than JSON today, except if a constraint requires it.

The `laminas-json-server` component enables you to expose PHP classes as services. Using PHP DocBlocks, the `laminas-json-server` component allows remote procedure calls to PHP functions and methods.

> **JSON-RPC reference**
>
> You can find the JSON-RPC specification at `https://www.jsonrpc.org/specification`.

LDAP

The **Lightweight Directory Access Protocol** (**LDAP**) is an open standard protocol for managing directory services. There is an open source implementation of LDAP named OpenLDAP available in several Linux distributions. Many companies sell LDAP services as RHDS by Red Hat or Azure Active Directory by Microsoft. You can find LDAP implementations in most cloud providers.

It is useful to have a component that makes it easy to manage LDAP operations in PHP applications. This is exactly what `laminas-ldap` does.

Log

You should remember that we have used the `error_log` function several times to save messages to the web server log. In most cases, we have used this function to save messages about exceptions. However, there was a time when we used the `error_log` function to save information about a successful authentication. It doesn't seem very appropriate to use a function with the word *error* in its name to save information messages, do you agree?

To manage log messages with appropriate identification and store the messages in a variety of backends, you can use the `laminas-log` component. This component conforms to the PSR-3 specification, available at `https://www.php-fig.org/psr/psr-3`.

There are other Laminas components too, but the ones that have been presented in this section are of general interest.

In the next section, we will talk about a component that is very useful for distributed architectures.

Microservice-oriented development

There is a special component, or better, a set of Laminas components, that can be considered a microframework or a framework for microservices. We are talking about **Mezzio**.

Mezzio was designed to implement the concept of middleware according to the PHP-FIG. In the PSR-15 specification, middleware "is an individual component participating, often together with other middleware components, in the processing of an incoming request and the creation of a resulting response" (`https://www.php-fig.org/psr/psr-15`). With Mezzio, you can easily create middleware pipelines to process HTTP requests.

Mezzio is very suitable for building reduced applications. This microframework basically uses four components for creating middleware: a router, a dependency injection container, a templating engine, and an error handler. By using minimal resources, Mezzio helps developers to create microservices in PHP.

You can find specific documentation about Mezzio at `https://docs.mezzio.dev`.

In the next section, we will talk about building APIs using Laminas components. In fact, we have already built APIs with Laminas components. The good news is that now we will reveal that there is a tool to make this easier.

Building APIs

The Laminas community provides a tool for building PHP APIs: **API Tools**. API Tools is implemented with Laminas components, but it does not require the developer to use Laminas as the framework. This means that you can use API Tools to generate APIs for PHP applications written with any library or framework. You can find documentation about API Tools at `https://api-tools.getlaminas.org`.

In the next section, we will present some useful Laminas resources provided by the community.

Community resources

In addition to the Laminas documentation, there are two other sources for learning about Laminas and asking questions:

- **Laminas forums**: You don't need to have an account at `https://discourse.laminas.dev` to read the posts on official Laminas forums. However, you need to create an account to post questions.

- **Laminas GitHub repositories**: The open source code of Laminas components is also documentation. You can find the repositories of each Laminas component at `https://github.com/laminas`. This web page has essential guidelines for getting support, contributing to the project, and reporting security issues.

In addition, you can find some examples of Laminas component extensions here: `https://github.com/fgsl/framework`.

Summary

In this chapter, we had a brief look at some Laminas components that were not used in the example application in this book. We learned about some components related to general purposes, such as the cache and logs.

We also learned that there is a specific component for creating distributed and decentralized applications, the Mezzio microframework. In addition, we have learned that the Laminas community provides API Tools for creating PHP APIs.

Finally, we learned about where to get help with Laminas and how to become a part of the Laminas community and improve the framework.

A real framework focuses on general issues. But you should take advantage of the extensibility and the open source features of Laminas to implement solutions for specific issues.

In conclusion, it is good to remember one of the principles of the Agile Manifesto: *"Simplicity – the art of maximizing the amount of work not done – is essential."* Use Laminas and simplify your work.

Create. Only what exists can be improved.

Index

Symbols

30-second rule 312

A

access control
  implementing 259-262
Access Control List (ACL) approach 259
Adapter design pattern 59, 76, 222
addUserAndRole function 246
agile development
  principles 310, 311
agile manifesto 310
  URL 50
Apache HTTP Server 14
Apache Web Server
  reference link 18
API
  building 362
  documentation, generating 321-328
API Tools 362
  reference link 362
architecture 4
Asynchronous JavaScript and
    XML (AJAX) 351
authentication, Inventory and Store modules
  generic behavior, decoupling 305-308

authentication listener
  delegating, to Inventory module 238-240

B

BDD, using with Behat 124, 125
  first scenario, creating 126, 127
  user story, mapping to automated
    tests 127-131
Behat 124
  BDD, using with 124
  models, creating with 146-155
business class 53

C

cache 357
clean code 312
code coverage reports
  generating 71-73
command-line interface (CLI) 358
Composer 15
  used, for PHPUnit installation 45-47
  using 48, 49
composer.json
  editing, directly 94
config.php file 59

container
Laminas, using as 92
context class 127
continuous delivery 4, 5
continuous integration 4, 5
controller actions
finishing 354
controller classes
refactoring 332, 333
controller layer
AJAX requests, detecting 350, 351
mastering 350
messages, exchanging between
view layers and 352-354
Create, Read, Update, Delete (CRUD) 196
current test cases, testing 160-164
implementing, with controller
and view pages 160
interaction, implementing between
web interface and API 182-188
interaction, testing between web
interface and API 188-192
pages, implementing for
ProductController 178-182
ProductAPIController class,
creating 170-173
ProductController class,
implementing 173-178
product deletion, testing 169, 170
product insertion, testing 164-166
product recovery, testing 167
product updates, testing 168, 169
Cross-Site Request Forgery (CSRF) 358
Cross-Site Scripting (CSS) 359
crypt 358
Cucumber
reference link 127

customer
controlling 297-30
home page, refactoring 287-293
customized filter
creating 338, 339
customized validator
creating 340-342

D

delete.php file
creating 70, 71
Document Object Model (DOM) 358
duplicated code
eliminating 311-315

E

Eclipse 19
Eclipse package, for PHP Developers
reference link 19
Eclipse PHP Development Tools (PDT) 19
Eclipse toolbar
Laminas, integrating into 21-29
edit.php
altering 68-70
employee's registration
creating 226-228
login page, testing 228-232
Escaper 359

F

feed 359
filters
customizing 337
form, for discounts
generating, with laminas-form 196

form view plugin 204
frameworks
    usage 6
Fully Qualified Class Name (FQCN) 353

G

generic model
    creating 142-145
getConfig function 245
getResources function 247, 248
Gherkin 127
grantPermissionsToRole function 248

H

HTML form
    defining, with class 200-205
hydrator 360
Hypertext Transfer Protocol (HTTP)
    versus PHP 157

I

identity document number (IDN) 302
identity management
    class, modifying 330-332
    controller classes, refactoring 332, 333
    isolating, for each module 329, 330
    listener classes, refactoring 332, 333
identity manager
    creating 267-274
input data 211
    filtering 205-208
    validating 208-212
integrated development environment (IDE)
    configuring 18-21

Inventory module
    authentication listener,
        delegating to 238-240
    login and logout actions, creating 214
    user authentication, implementing 220-225
inversion of control pattern 159

J

Java Development Kit (JDK) 19
JSON responses
    producing, with laminas-view 343, 344
JSON-RPC server 360

L

Laminas
    installing 15
    installing, through isolated component 16
    installing, through MVC skeleton
        application 16-18
    integrating, into Eclipse toolbar 21-29
    reference link 10, 18
    technical and social infrastructure 10, 11
    using, as container 92
    versus Zend Framework 10
    Zend Framework, migrating to 6, 7
Laminas, community resources
    forums 362
    GitHub repositories 362
Laminas, using as container
    school3 project, creating 92, 93
    School module, creating 93, 95
    School module namespace, ensuring 95
    test methods, implementing 96
    tests, running 97-107

Laminas components

cache 357

command-line interface (CLI) 358

crypt 358

CSRF 358

DOM 358

Escaper 359

feed 359

hydrator 360

JSON-RPC server 360

laminas-permissions-acl component 259

laminas-permissions-rbac component 259

Lightweight Directory Access
        Protocol (LDAP) 360

log 361

laminas-dom

development state 359

laminas-form 199

CRUD implementation,
        creating with 196-198

files, uploading from 345-348

form for discounts, generating with 196

Laminas forums

reference link 362

Laminas GitHub repositories

reference link 362

Laminas MVC application

request lifecycle 158, 160

testing, errors in PHP 7 93

laminas-view

JSON responses, producing with 343, 344

Laminas view layer 342, 343

application layout, changing 348, 349

**Lightweight Directory Access
        Protocol (LDAP) 360**

reference link 361

listener classes

refactoring 332, 333

loadAuthorizationData method

defining 261

log 361

login and logout actions

creating 215-220

creating, for Inventory module 214

tests, creating 214, 215

login page

testing 228-232

M

Manifesto for Software Craftsmanship 310

mapped models

creating 336

mappers

creating, from user stories 135

Mezzio 11, 361

reference link 361

microservice-oriented development 361

model-entity relationship (MER) 115

models

creating, from user stories 131-134

Model-View-Controller (MVC) 9

MVC skeleton application

Laminas, installing through 16-18

MySQL databases

managing 29-34

MySQL/MariaDB 14

O

object-oriented programming (OOP) 3, 42

object-relational mapping (ORM) 16

using 76, 78

OpenAPI Specification 321

OpenLDAP

reference link 361

Open Web Application Security
    Project (OWASP) 359

P

perspective 21
PHP
    reference link 358
    versus HTTP 157
PHP 7
    Laminas MVC, testing errors 93
PHP Copy/Paste Detector (PHPCPD) 311
PHP files
    purposes 41, 42
PHP Framework Interoperability
    Group (PHP-FIG) 8
PHP Mess Detector (PHPMD) 315
phpMyAdmin 15
PHP Standard Recommendations (PSRs) 8
PHPUnit
    installing, Composer used 45-49
PHPUnit tab 54
ProductAPIController class
    creating 170-173
product basket
    controlling 293-297
ProductController class
    implementing 173-178
ProductControllers
    pages, implementing 178-182
product inventory
    managing 280-287
products
    adding, to products table 136-140
    insertion, checking 140, 141
product search box
    creating 318-320

public attributes
    using, in class 63
purchase order page
    code, for closing 301-304

R

Recover and Update (RU) 280
recreateDatabase function 245
resources registration
    creating 243
resources table
    filling 244-250
Role-Based Access Control (RBAC) 241
role resources
    implementing 250-254
roles registration
    creating 242, 243
Row Data Gateway 76, 77
RSS
    documentation link 360

S

save.php file
    altering 70
school1 project
    creating 39, 40
    implementing 41
SchoolClass class
    attributes 42
school classes
    automated test, creating 45
    managing 38, 39
    structure, implementing 39, 40
SchoolClass::getAll method 59
SchoolClass.php 55

school class registration

implementing 42-45

SchoolClassTest.php 52, 53

Singleton pattern 80, 81

software requirements 38

students management use case 78

CRUD files, adjusting 87-92

implementing 78, 79

SchoolClassTest->testListing
test, running 80-87

Swagger

reference link 321

Swagger Specification 321

T

Table Data Gateway pattern 76, 77, 135, 136

Technical Steering Committee (TSC) 10

test bootstrap

modifying 265, 266

test-driven development (TDD) 41, 214, 310

U

**Unified Modeling Language (UML)
component diagram 41**

Uniform Resource Locator (URL) 32

unnecessary configuration

removing 316-318

use cases, developing with TDD

configuring, autoloading of classes 55-57

configuring, autoloading of test classes 57

config.php file, creating 58-60

delete.php file, creating 70, 71

developing 49

edition form edit.php, creating 63, 64

edit.php, altering 68-70

save.php, altering 70

SchoolClass class, creating 54, 55

SchoolClassTest->testUpdating test
method, creating 65-68

SchoolClassTest test case, rerunning 58-60

SchoolClassTest test class, creating 50-52

testInserting test method, creating 60-63

test method, creating in
SchoolClass class 53, 54

user authentication

implementing, for Inventory
module 220-225

user identity

verifying, based on events 232-238

user role

implementing 254-259

users permissions

verifying 263-265

user stories

mappers, creating from 135

models, creating from 131-134

V

validator class 212

validators

customizing 337

variable function 286

view 21

Visual Studio Code (VS Code) 18

W

web frameworks

need for 3, 4

web interface, and API

interaction, implementing 182-188

interaction, testing 188-192

Whatstore virtual store
  actors 113-115
  class diagram 115, 116
  Laminas project structure 117, 118
  modules, preparing 118-121
  project instance, creating 116
  project requirements 112
  use cases 113-115

X

XAMPP 14
XMLHttpRequest
  URL 351
xUnit pattern 124

Z

Zend Framework 7, 8
  migrating, to Laminas 6, 7
  reference architecture, for PHP
    applications 8-10
  versus Laminas 10

`Packt.com`

Subscribe to our online digital library for full access to over 7,000 books and videos, as well as industry leading tools to help you plan your personal development and advance your career. For more information, please visit our website.

Why subscribe?

- Spend less time learning and more time coding with practical eBooks and Videos from over 4,000 industry professionals

- Improve your learning with Skill Plans built especially for you

- Get a free eBook or video every month

- Fully searchable for easy access to vital information

- Copy and paste, print, and bookmark content

Did you know that Packt offers eBook versions of every book published, with PDF and ePub files available? You can upgrade to the eBook version at `packt.com` and as a print book customer, you are entitled to a discount on the eBook copy. Get in touch with us at `customercare@packtpub.com` for more details.

At `www.packt.com`, you can also read a collection of free technical articles, sign up for a range of free newsletters, and receive exclusive discounts and offers on Packt books and eBooks.

Other Books You May Enjoy

If you enjoyed this book, you may be interested in these other books by Packt:

Clean Code in PHP

Carsten Windler, Alexandre Daubois

ISBN: 978-1-80461-387-0

- Build a solid foundation in clean coding to craft human-readable code
- Understand metrics to determine the quality of your code
- Get to grips with the basics of automated tests
- Implement continuous integration for your PHP applications
- Get an overview of software design patterns to help you write reusable code
- Gain an understanding of coding guidelines and practices for working in teams

Rust Web Development with Rocket

Karuna Murti

ISBN: 978-1-80056-130-4

- Master the basics of Rust, such as its syntax, packages, and tools

- Get to grips with Rocket's tooling and ecosystem

- Extend your Rocket applications using Rust and third-party libraries

- Create a full-fledged web app with Rocket that handles user content

- Write pattern-matching logic and handle Rust object lifetimes

- Use APIs and async programming to make your apps secure and reliable

- Test your Rocket application and deploy it to production

- Containerize and scale your applications for maximum efficiency

Packt is searching for authors like you

If you're interested in becoming an author for Packt, please visit authors.packtpub.com and apply today. We have worked with thousands of developers and tech professionals, just like you, to help them share their insight with the global tech community. You can make a general application, apply for a specific hot topic that we are recruiting an author for, or submit your own idea.

Share Your Thoughts

Hi!

I am Flávio Lisboa, author of *PHP Web Development with Laminas*. I really hope you enjoyed reading this book and found it useful for increasing your productivity and efficiency in Laminas.

It would really help me (and other potential readers!) if you could leave a review on Amazon sharing your thoughts on *PHP Web Development with Laminas*.

Go to the link below or scan the QR code to leave your review:

https://packt.link/r/1803245360

Your review will help us to understand what's worked well in this book, and what could be improved upon for future editions, so it really is appreciated.

Best Wishes,

Flávio Lisboa

Download a free PDF copy of this book

Thanks for purchasing this book!

Do you like to read on the go but are unable to carry your print books everywhere? Is your eBook purchase not compatible with the device of your choice?

Don't worry, now with every Packt book you get a DRM-free PDF version of that book at no cost.

Read anywhere, any place, on any device. Search, copy, and paste code from your favorite technical books directly into your application.

The perks don't stop there, you can get exclusive access to discounts, newsletters, and great free content in your inbox daily

Follow these simple steps to get the benefits:

1. Scan the QR code or visit the link below

https://packt.link/free-ebook/9781803245362

2. Submit your proof of purchase
3. That's it! We'll send your free PDF and other benefits to your email directly

www.ingramcontent.com/pod-product-compliance
Lightning Source LLC
Chambersburg PA
CBHW062037050326
40690CB00016B/2969